Circle of the Seasons

Books by Edwin Way Teale

GRASSROOT JUNGLES
THE GOLDEN THRONG
NEAR HORIZONS
DUNE BOY
THE LOST WOODS
DAYS WITHOUT TIME
NORTH WITH THE SPRING
JOURNEY INTO SUMMER
AUTUMN ACROSS AMERICA
WANDERING THROUGH WINTER
A NATURALIST BUYS AN OLD FARM
A WALK THROUGH THE YEAR
SPRINGTIME IN BRITAIN
ADVENTURES IN NATURE
THE JUNIOR BOOK OF INSECTS
THE STRANGE LIVES OF FAMILIAR INSECTS
PHOTOGRAPHS OF AMERICAN NATURE

Books Edited by Edwin Way Teale

WALDEN
GREEN TREASURY
AUDUBON'S WILDLIFE
THE THOUGHTS OF THOREAU
THE INSECT WORLD OF J. HENRI FABRE
THE WILDERNESS WORLD OF JOHN MUIR

EDWIN WAY TEALE

Circle of the Seasons

The Journal of a Naturalist's Year

The Edwin Way Teale Library of Nature Classics

DODD, MEAD & COMPANY • NEW YORK

Dedicated to

DAVID M. EDWARDS

a friend at the crossroads

A few of the entries in this journal were originally published in *The Bulletin of the Brooklyn Entomological Society, The Long Island Naturalist, Natural History Magazine,* or through the George Matthew Adams Service.

Reissued 1987 by Dodd, Mead & Company, Inc.
71 Fifth Avenue, New York, N.Y. 10003
Manufactured in the United States of America
Designed by Linda Huber
1 2 3 4 5 6 7 8 9 10

Library of Congress Cataloging-in-Publication Data

Teale, Edwin Way, 1899–1980
Circle of the seasons.

(The Edwin Way Teale library of nature classics)
Includes index.
1. Natural history—Outdoor books. 2. Seasons.
I. Title. II. Series.
QH81.T274 1987 508 86-24286

ISBN 0-396-09015-X

ISBN 0-396-09016-8 {PBK}

Contents

Circle of the Seasons

CHAPTER ONE

January

JANUARY • 1

SONG OF THE WHITETHROAT Windless, silent, under a low ceiling of gray, this first morning of the new year is like an echoing room. Sounds carry far as I walk the mile along the swamp edge, past the Insect Garden hillside and on to Milburn Pond. In a tangle of cat-briar and shadbush, near the edge of the frozen water, a white-throated sparrow is singing a snatch of its springtime song. Again and again, I hear the pure, ethereal strain, simple, moving, bringing back in memory a late spring day on a lake shore in the forests of Maine. No other voice among all the singers of nature affects me more deeply. The song of the white-throated sparrow—how fine a beginning for a new year!

JANUARY • 2

TOMORROW Whatever tomorrow brings, the grass and the trees and the birds will see it.

MOONLIGHT Walking tonight under a clear sky, I observe the bare roofs frosted with the rime of moonlight.

JANUARY • 3

ACORN IN A COCOON Toward sunset today, I stop beside the tangle of cat-briars where the whitethroat sang. A weathered, last year's cecropia cocoon is attached to one of the briars. The upper end shows a round opening where the body of the moth squeezed its way to freedom many months before. I touch the cocoon. Within, something rattles. I peer through the opening. An acorn is tucked inside. I have chanced upon the secret cache of a bluejay.

JANUARY • 4

MIDWINTER The commonest simile in connection with the new year is a book with blank pages. Nature's year is also a book to be written. This midwintertime represents a pause in the turning wheel of life. It is, in northern lands, the year's low point, its nadir. Life will swell, reach its zenith, before the next resting time. All the events of spring and summer and autumn, of sprouting and growth and seed time, the beginning and the end, lie ahead. The whole circle of the seasons stretches away before us as we view the year from the cold plateau of January.

JANUARY • 5

A CAT KNOWS MY FOOTSTEPS Every morning, when I walk the better part of a mile for the papers, I pass a house where a gray and white cat lives. For some reason it considers me a special friend. Ever

since it was a kitten, it has been coming out to meet me, walki
hundred feet or so with me, rubbing against my legs when I stop, purring loudly when I stroke it or scratch its head. I go by its house each morning at about 7:30, but the hour varies from day to day. But each time, as I come down the walk, often when I am a considerable distance from the house, the cat comes running to meet me. Frequently it comes from behind the house as I approach. It cannot see me. Often the wind is from the wrong direction. It cannot smell me. It recognizes the sound of my footsteps. Thus it tells me from all others passing by.

JANUARY • 6

TIME AND TRUTH You can prove almost anything with the evidence of a small enough segment of time. How often, in any search for truth, the answer of the minute is positive, the answer of the hour qualified, the answers of the year contradictory!

JANUARY • 7

THE WINTER COLD Protected by sweaters and a leather jacket against the biting blasts of the north wind, I walk along the hillside this afternoon. Snow lies drifted among the wild cherries. Where the wind has swept bare the ground, the soil is frozen and rocklike. On this day of bleak cold, the earth seems dead. Yet every northern field and hillside, like a child, has the seeds and powers of growth locked within it. From cocoon to bur, on a winter's day, there is everywhere life, dormant but waiting.

Within the earth there are roots and seeds; on the bare twigs, there are winter buds; buried in soil and mud beneath ice-locked water are the turtles and frogs and dragonfly nymphs; hidden in decaying logs and under snowcovered debris are the fertilized queens of the wasps and bumblebees. Everywhere, on all sides of us, as far as winter reigns, life is suspended temporarily. But it has not succumbed. It is merely dormant for the time being, merely waiting for the magic touch of spring. All the blooms of another summer, all the unfolding myriad leaves, all the lush green carpet of the grass, all the perfumes of the midsummer dusk, all the rush and glitter of the dragonfly's wings under the August sun—all these are inherent, locked up in the winter earth.

Nor is this time of suspended activity wholly wasted. Scientists have discovered that, for many kinds of seeds, a period of cold is essential to their proper sprouting. They require the months of cold just as they do the days of spring.

Seeds that lie on the frozen ground, that are coated with sleet and buried by snow, are thus the most favored of all. Bring those same seeds indoors, coddle them, keep them warm, protect them from wind and cold and snow, and they sprout less readily in the spring. The seeming punishment of winter is providing, in reality, invaluable aid. Similarly, the eggs of some insects, such as the Rocky Mountain locust, need cold for proper hatching.

Winter cold, the enemy of the easy life, thus is not the enemy of all life. It aids in the proper development of seed and egg. The deathlike inactivity of the winter earth is only an illusion. Life is everywhere, in every foot of frozen soil, in every rocklike yard of solid ground—life in the endless variety of its natural forms.

JANUARY • 8

OWL ON A CORN SHOCK I have just heard of the unusual daytime activity of a great horned owl last fall, near Orient, at the far end of Long Island. It appeared in a field where a farmer was husking corn. Paying little attention to the man, it perched on the top of one of the frosty corn shocks. As he worked, the farmer moved from corn shock to corn shock, overturning each before beginning the husking. Meadow mice and house mice, living snug under the shocks, thus finding themselves dispossessed, poured out into the field. As each shock went over, the owl swooped down, snatched up scurrying rodents and then returned to its perch. Thus, for several hours, it fed well and rapidly reduced the rodent population of the corn field.

JANUARY • 9

THE NOBLE RED MAN The "Noble Red Man" never was and never is—it is not races but individuals that are noble or courageous or ignoble and craven or considerate or persistent or philosophical or reasonable. The race gets credit when the percentage of noble individuals is high.

JANUARY • 10

WREN NESTS Today I have been making a harvest of long-billed wren nests in the frozen swamp. These oval balls of twisted vegetation, solidly attached to upright cattail stems, form the cold-weather homes of varied creatures—small beetles, overwintering wasps, bumblebees and spiders.

The whole swamp edge this morning was embroidered with the white lace of frost and ice. Here and there, like drooping petals, tongues of ice issued from holes in the marsh muck, forced up by the pressure of expansion as the ground moisture congealed in a sudden freeze. The cattail swamp spread away in a ragged plain of yellow-gray at the foot of the snow-clad hill.

As I push my way through the cattail stands, walking on solid ice, I am surrounded by a moving cloud of fluff. The heads are ripe, swollen, ready to send off the winged seeds at a touch. I zigzag this way and that, hunting the nests. The cattail fluff gets into my nose, my eyes, my mouth. My hat and clothes are covered with it. But, in the end, I bring home half a dozen nests. This year I find the majority of the winter sleepers snug in these hibernacula are pale, straw-colored little spiders.

JANUARY • 11

POISON IVY BERRIES To the cedar woods at sunset. An overwintering catbird is feeding on the purple berries of the cat-briars and the gray berries of the poison ivy. Cows eat poison ivy leaves with immunity; birds eat the berries without ill effects. Only human beings, and only some human beings at that, are susceptible, through an allergy, to the torments of ivy poisoning. Yet those of us who fall victims, through this peculiarity in our own systems, are apt to bewail the presence of poison ivy in the world and to consider it an enemy lying in wait for our approach. Why was poison ivy put in the world?

The catbird, feeding in the quiet of this winter sunset, finding nourishment in the berries, supplies at least one of the answers. On the shifting sand of Fire Island, there is another answer. There, the remarkably adaptive ivy, able to grow on barren sand as well as in swamps, anchors down the windblown earth. Each form of life, plant as well as animal, has its part to play. Whether the results delight or outrage man is no concern of nature's. Man plays, to be sure, a leading role in the drama of life. But he still is only a part of a cast that includes many, many actors.

JANUARY · 12

WINTER WATER This morning, when the sun began to melt a heavy frost on a garage next door, I watched the starlings drinking drops of water that collected at the bottom edge of the composition shingles. These birds are particularly resourceful in getting water. One summer on Staten Island, during a prolonged drought, the starlings were seen walking along heavy telephone cables, reaching underneath and drinking small droplets of moisture that had collected there during the night.

It is in midwinter that birds are sometimes hardest put of all to find sufficient water. Eating snow provides some moisture and birds, such as the cedar waxwings, sometimes catch snowflakes as they fall through the air. But when the ground is bare and water frozen, moisture is at a premium for many species.

A few days ago, a young bluejay landed on the dry golden-glow plant that curves over in an arc near our kitchen door. The previous night, a sudden fall in temperature after a mild spell had frozen drops of water into little clear beads of ice on the underside of the curving stalk. Over and over again, the bluejay opened its bill, stretched out its neck and tried to drink the frozen drops. They seemed bewitched. The bird peered closely at the congealed water and tried again. But no fluid ran down its throat. It seemed completely bewildered and finally flew away after its first experience with undrinkable winter water.

JANUARY · 13

OPTIMISM Optimism is more likely if we keep our eyes on the average, the general, rather than the exceptions, the individual. The bat caught on burdock burs—this is the exception, the tragic, the unusual event. It is best not to dwell too long on the exceptions. But what if we are exceptions?

JANUARY · 14

SOUNDS OF WINTER Toward the end of this still, piercingly cold winter's day, I crossed the snow-covered field beside the swamp and turned into a winding road. At every step, the snow squeaked shrilly beneath my feet. Distant sounds carried with startling distinctness. I heard the barking of a dog, an automobile engine starting, the mellow whistle of a locomotive far away. As I walked on, I fell to wondering about winter sounds.

Why do sounds seem to carry farther through cold air? Why do they reach the ear more distinctly in winter than in summer? I remember a farmer in Indiana who claimed he could guess where the mercury stood in the thermometer on clear midwinter days merely by listening to the sound of car wheels on a railroad track half a mile away beyond his pasture. The colder the day, the more distinctly every metallic rattle came to his ears. And why is it that snow squeaks loudest on the coldest days? I now know the answers to these questions because I have just put them to a scientist friend of mine. Here are his explanations:

The squeaking of the snow is produced by sharp edges of frozen crystals rubbing against each other. In mild weather, when the mercury stands just a little below freezing, the edges are less hard, more easily blunted. But as the mercury descends, the crystals become correspondingly harder, friction mounts and the high-pitched complaint of the rubbing corners and edges increases. "Snowball snow" that packs easily is squeakless. It is the harder crystals of the colder days that produce the shrill fifing of the snow beneath our feet.

The drop of the mercury is also accompanied by the condensing of the atmosphere. When air is warmed, it expands and the molecules separate; when it is cooled, it contracts and the molecules draw closer together. This is the secret of far-carrying sounds on the winter air. The denser the medium, the better it transmits sound waves. If someone taps two stones together fifty feet away in the air, the sound is hardly noticeable. But if he does the same thing underwater, while you have your head beneath the surface, the sound is so penetrating it is painful. Water, denser than air, is a better conductor of sound. Steel rails are even better. Put your ear to a rail and you can catch sounds of an approaching train long before any noise reaches you by air.

If this is true, then why do we not hear sounds clearly in a blizzard when the density of the atmosphere is increased by falling flakes of snow? Instead, sounds are "blanketed" by snowfall. This is so, my friend explains, because the sound waves strike the solid matter of the flakes and are reflected back. The multitude of flakes in a blizzard cut out a large proportion of the waves. We hear only part of the sound, that part that passes between the falling flakes, and thus it reaches us in diminished volume.

JANUARY • 15

CAT IN A CROWD This day, I went to the city. And the most interesting thing I saw there was a cat.

I first noticed it as I was hurrying through the Pennsylvania Station. I stopped hurrying. Gray and white, and about half grown, the station cat sat in the middle of the thronging Long Island concourse. Unconcernedly, it watched hundreds of people streaming by. Like a little island in a river, it broke the flow of human beings; those on the right swerving around it, hurrying in one direction; those on the left, hurrying in the other direction, doing likewise. A man, arguing with his wife, almost trod on its outstretched tail. At the last instant, a nonchalant twitch of the tail flipped it from under the descending foot. A Negro, pushing two heavy trucks of baker's tins, deployed around the cat. A nearsighted woman bent down to peer at it. An elderly man scratched its head. The cat got up and rubbed against his legs. Then it sat down again.

Three women approached in animated conversation. They stopped talking and walking to stare at the cat. Then, talking again, they moved on. A child patted its head and a messenger boy, delivering three dish mops, came along, stroked the cat with one of the mops and sauntered away. Trains came and went. The crowds swelled and diminished. The station cat seemed used to having a torrent of human beings pouring around him. The last I saw of him he was walking nonchalantly across to one of the metal posts at the side of the concourse, lifting his forepaws high and stretching—a city cat trying to sharpen his claws on a metal tree.

JANUARY • 16

A SPARROWHAWK LOSES A STARLING About 3:45 on this windy afternoon, there was a sudden commotion near a weeping mulberry tree in a neighboring yard. Two birds were struggling on the ground. At first I thought they were fighting. Then, through glasses, I saw a sparrowhawk was gripping a starling. The captured bird, almost as big of body as its captor, was putting up a great resistance, pecking with its long bill at the little falcon that stood above it with wings downspread.

As I drew near, the hawk took flight, alighting in a neighboring pear tree, while the starling hurled itself pell-mell into the maze of drooping branches of the weeping mulberry tree. There it was safe and there it stayed. Nothing could dislodge it. Two chickadees within the same tree kept up an incessant calling and scolding. All the sparrows had disappeared from the yard.

Hardly had I moved away before the hawk flew down to a clothesline, balancing itself with difficulty in the gusts, spreading its brilliant rufous tail as it bobbed on its unstable support. From there it glided to the top of a wire fence where it turned first to one side, then to the other, peering down at the ground on either hand. Then it dropped to the yard, alighting at the exact spot where it had lost the starling. Moving about in awkward little jumps, it searched this way and that over the ground for several minutes. There it had lost its prey and there it seemed sure it would find it again.

Perhaps five minutes went by before it gave up the search and flew away. The starling, apparently little injured, waited for a long time after it was gone before emerging from the shelter of the mulberry tree. This was the first time I ever saw a sparrowhawk capture a bird so large and the first time I ever saw one of these little falcons moving about over the ground in search of a bird it had lost.

JANUARY • 17

WOOLLY BEARS Down the side of the hill and across the path at the swamp edge runs a strip of russet moss like an outcropping vein of metal. Above this darker moss, the snow has melted first on this day of milder weather. Here I roll over a fragment of a log and find beneath three hibernating woolly bears.

The caterpillars are tightly curled, like little black and red-brown doughnuts. They show no sign of life. All across this northern land, on this winter day, woolly bears are lost in their long sleep, curled up beneath old boards, buried snugly beneath discarded rubbish, protected from wind and snow, dozing the winter away.

Each furry, hibernating caterpillar is motionless, its position unchanging for months on end. Uncover it, place it in your hand and hold it there until it is warmed and its apparent lifelessness disappears. It straightens out, stretching, so to speak, after its long slumber. Then it begins to walk about. The temporary warmth has brought the awakening which normally is produced by the coming of spring. Lay it down again in its protected position and it will soon curl up like a sleepy kitten and fall once more into the profound slumber of its winter hibernation.

In the autumn, we see these woolly bears humping along, hurrying across highways and open spaces, searching for the best places in which to retire for their long sleep. Sometimes they stop and stretch out the front of their furry bodies, swinging from side to side as though peering nearsightedly about them. And they certainly are poorly equipped in the matter of eyes. I have to look through a pocket magnifying glass to see the woolly bear's eyes, so tiny are they.

It may be that the bristly hairs that cover its body provide the caterpillar with sensory outposts that make keen vision unnecessary. At any rate, these hairs, by their resiliency, enable a woolly bear to slip easily from between our fingers. They also make the caterpillars unappetizing to many birds which, in consequence, leave them off their bill of fare.

When the winter sleep is over and spring comes, the bristles help in still another way. In April or May, each of the woolly bears now slumbering beneath the log will weave a little cocoon of silk. Hairs shed from its body will be woven into this covering until the completed case has the appearance of being formed of felt. It is within this cocoon that the woolly bear will leave behind its larval shape and be transformed into the winged adult Isabella Tiger moth. The delicate beauty of this little moth is referred to by John Keats in the lines: "Innumerable of stains, with splendid dyes, as are the Tiger Moth's deep damask wings."

The forewings of this moth are tawny and the hind wings often stained with orange-red. Three rows of black dots, six in each row, decorate its reddish body. One row runs along its back, the other two are on either side of its body, like lines of tiny darkened portholes. The beauty of this adult woolly bear is rarely seen, however, as it is active only at night. From eggs laid by the females, a new generation of woolly bear caterpillars begin feeding and growing, preparing for the time of long slumber when another winter comes.

JANUARY · 18

BLUEJAY THRIFT There is, in this neighborhood, a tame bluejay that has the run of the house. It was rescued, a bedraggled fledgling, after it was thrown from the nest by a summer storm. This bird's favorite food is puffed rice. True to bluejay instinct, it often stores away part of its meal, hiding kernels of the breakfast food between slats of the pulled-up Venetian blinds. Whenever the blinds are lowered, a white blizzard of puffed rice descends with them.

JANUARY · 19

DEATH Why death? Fundamentally—in nature's scheme of things —it is a means of utilizing food, that an everlasting present would consume, for nourishing better forms of life. If all the fish in the sea, all the earthworms in the soil, all the deer in the forest, lived forever, evolution would come to a standstill. The globe would reach a saturation point where improved forms of life would be impossible. Ours is a world of unabating change. Only by making way for new individuals can evolution's endless chain continue.

JANUARY · 20

A SQUIRREL HIDES PEANUT BUTTER Before breakfast each morning, my wife stocks the bird-feeder that hangs from the limb of a backyard maple with sunflower seeds and peanut butter for the chickadees and nuthatches. Today, just as she came in from this task, Nellie noticed that one of the gray squirrels had caught the scent of the peanut butter. It clung to the trunk, pointed its nose in the direction of the hanging feeder, and leaned far out into space. It climbed to the limb over the feeder and looked down. It tested the wire. Finally, it leaped from the trunk, caught the swaying feeder with its forepaws, scrambled up and thus reached the peanut butter contained in the cap of a Mason jar. It ate its fill. Then, before dropping to the ground, it carefully covered over the remaining nut butter with the empty shells of the sunflower seeds, hiding the cache as it might bury a nut in the ground.

JANUARY • 21

ZIGZAGS OF NATURE I stop beside an oak tree this afternoon to watch a white-breasted nuthatch hitching itself upward along the trunk. Moving first to this side, then to that, it ascends the tree in zigzags. Wild creatures often follow a crooked path. Ants hunt their food on plants by tacking and veering. Squirrels hop this way and that on the ground. The great night-flying moths, the Polyphemus, the Cecropia and the Luna, all follow a zigzag trail through the darkness. The ant, with myopic vision, thus finds food it otherwise might miss if it ran in straight lines across the leaves. The zigzagging of the squirrel enables it to see all around it, to be aware of its surroundings and to avoid surprise attacks when it is at a disadvantage on the ground. By weaving widely, the night moths increase their chances of catching the scent trails that the females leave on the night air. The zigzag path plays an important part in the lives of many creatures.

JANUARY • 22

SNOW CLOUDS About seven this morning, I awoke to a still, closed-in day. Lead-colored clouds hung low and the mercury stood at 28 degrees F. The smell of snow was in the air when I walked for the morning papers. The sky seemed weighted down, pressed low, with a myriad unborn flakes. A few fell slowly, turning and drifting down in the hushed, unmoving air. When the gray and white cat trotted to meet me, I noticed a dozen or more flakes caught on its fur. By the time I reached home, the snow was falling heavily.

Hour after hour, flake piled on flake. There was no wind, no noise, no violence, only the steady, increasing descent of the snow. The ground was soon covered and the backyard trees became gray and ghostly in the storm. The flakes sifted down between the branches of the trees, slid along the trunks, piled up around the rose bushes, covered the rock garden. The snow was falling at noon; it was falling at dusk; it was falling in the darkness when I went to bed. The last thing I saw as I looked out into the silent storm was the spread fingers of the spruce twigs clad in snow as though in white woolen gloves.

JANUARY · 23

WIND All night long, the snow continued to fall. Sometime before dawn, the wind rose. When I awoke, gray-white, scudding snow was streaming past my window. The gust-driven flakes scraped and scratched along the glass. Trees tossed in the murky light. I could hear their creaking and the rubbing of branches, sounds diminished and muffled by the curtaining snow.

It was afternoon before the gale blew itself out and the snow stopped falling. Drifts rose high along the fence, curved and twisted and shaped by the wind. Branches were down, some completely buried beneath the snow. Late in the afternoon, I floundered over to the swamp and along the hillside. Everywhere I looked, whichever way I turned, I saw, written on the snow, the autographs of the winter wind.

JANUARY · 24

THE WHITENESS OF THE SNOW Looking across the white fields today, I am reminded that an invariable rule of nature is that nothing is invariable. "As like as two peas in a pod" is an exact statement of nature's ways. For no two peas in a pod are ever exactly alike. Nature does not plagiarize herself, repeat herself. Her powers of innovation are boundless. No two hills or ants or oranges or sheep or snowflakes are identical. Ten thousand seem alike because we do not see them clearly enough, because our senses are too dull or inadequate or inaccurate to detect the differences.

And so it is with one of the oldest similes in the world: "As white

as the snow." The whiteness of the snow is infinitely varied.
its whiteness is produced by elements that are not white at a
individual crystals that go into the make-up of a snowflake are transparent and colorless when they are created far up in the sky. They are like clear glass. It is when they are grouped together in flakes, when they lie in untold millions in a drift, that they appear white.

What is the explanation of this paradox? It is the same answer that explains how a transparent window pane when it is broken and powdered appears white while the intact pane is colorless. Both the infinite number of crystals that make up the snowdrift and the vast number of particles that comprise the powdered glass have so many facets that they reflect all the rays of light in all directions. Put all the rays of the spectrum together and you have white just as when you take all the rays of the spectrum away you have black. It is the numberless crystals in the piled-up snow that turn it into a mound of the purest white.

Yet even this "purest white" has many subtle variations. The famous New York advertising photographer, H. I. Williams, once told me of his surprise in noting the differences in the whites recorded by his color camera. They all looked alike to his eyes. But the sensitive color film showed that some were tinged with blues or yellows or reds and some were pearly and opalescent. The white of a billiard ball, of a sheet of writing paper, of a dress shirt, of a tablecloth all were different. Probably no two tablecloths are exactly the same in whiteness, although our eyes are unable to detect the difference.

Similarly, it is likely that no two snowbanks are identical in the whiteness of their exterior. Their surroundings, the time of day, the conditions of the sky all contribute to their tinting. Even our eyes can note the blue in the shadows of trees stretching across the drifts and the pink glow of sunset spreading over an expanse of snow. But under the noonday sun, except when soot or mud has stained them or when old drifts have been discolored by deposits from the air, the whiteness of the snow remains all the same to our eyes—the purest white we know.

JANUARY · 25

ANTS IN A TREE An old maple tree, that I have known for many years near the Insect Garden, went down in the recent storm. I examined it today and found that all the interior was riddled with the tunnel maze of carpenter ants. Only a shell of wood and the bark of the exterior remained where the trunk had broken. The ants had mined away the strength of the tree. And, in so doing, they had destroyed themselves. They were exposed and homeless in the winter cold. The tunnels, where the break had occurred, were black with the massed bodies of the hibernating insects.

I brought home some of the ants and placed them in a glass beneath the warmth of my desk lamp. I was interested to observe that *the smallest ant revived first*. There is a sound basis for this occurrence. The smaller the creature, the greater the surface area in proportion to body bulk. It is this fact that explains why all the insects have to be cold-blooded. Heat would radiate away from the proportionately large surface areas of their bodies too rapidly for them to exist as warm-blooded animals.

JANUARY · 26

THE MOUSE IN A BIRD'S NEST In moderating weather, today, the snowdrifts are shrinking. Here, so close to the sea, the salt air hastens the melting of snow. Even in the cedar woods, the level of the drifts has dropped. I follow the trail around the edge of the wood, winding often through underbrush higher than my head. Three times, on this winter walk, I come upon deserted catbird nests that have been filled in and heaped high with grass and shredded bark. Each is the penthouse home of a mouse.

When I touch the mounds with an exploring finger, out comes one of the most appealing of small creatures. Its fur is silky, its ears large and delicate, its feet and the underside of its body immaculately white. Only half alarmed, it regards me with dark and liquid eyes and then runs nimbly down the bushes to the ground.

Called by many names—the deer mouse, the white-footed mouse, the wood mouse, the vesper mouse—this little creature has all the charm of a Walt Disney creation come to life. It is friendly and easily tamed. And it is so clean that the baby mice, even before they have their eyes open, will wash their faces carefully after every meal. As many as four litters of young a year keep the deer mouse population at a high level in spite of the work of owls and hawks, snakes and weasels. No other mouse, with the exception of the familiar meadow mouse, is so numerous or so widespread.

Years ago, when Nellie and I lived for a time in a trapper's cabin in the Maine woods, the first thing I saw when I swung the door open on its rusty hinges was a family of friendly deer mice. They were all sitting up on the edge of the bunk, without the slightest sign of fear, watching me with interest.

In varied species, deer mice are found from the Arctic Circle southward. They live below sea level in Death Valley and on mountains as high as there is vegetation to provide food. Their nests may be in hollow logs, under rocks, in deserted flicker holes, as well as in the abandoned homes of bush-nesting birds. Mainly nocturnal, they remain active all winter without hibernating. Like chipmunks and red squirrels, they store up seeds and nuts.

Besides their appealing appearance, their friendly ways, their cleanliness, these small creatures of the out-of-doors have an additional attraction. They are, in some species, musical mice. Occasionally, one of these dwellers in a second-hand home will lift its voice in a quavering little song that sounds not too unlike the high-pitched trilling of a canary.

JANUARY • 27

WINTERGREEN BERRIES The snow is gone; the ice has melted from the swamp stream. In a sudden January thaw, the thermometer has risen to 45 degrees F. On the brown ooze of the stream, I see the crisscrossing leaves of the fallen sweet flags and, near the russet moss, a new mole tunnel runs for a dozen paces along the hillside.

Under fallen oak leaves in the woods, I uncover a little clump of wintergreen. Here it is winterpink instead of wintergreen. All the leaves have turned a pinkish-red. Beneath a few of the leathery leaves, I come upon the brilliant red of the berries. They are not the plump wintergreen berries of the Lone Oak woods, remembered from boyhood in the dune country of northern Indiana. But they are filled with flavor—flavor that provides sure and swift transport to other days. How many years drop away at the taste of a wintergreen berry!

JANUARY • 28

SPRING IN JANUARY The thermometer, on this abnormal day of spring in January, stands at 60 degrees in the shade. A snowdrop is in bloom. Children are going fishing. Spiders are out running over the fallen leaves and the persisting green of the winter mosses is greener than ever today. Along the edge of the muddy swamp stream, little fishes are massing in the shallows. For a dozen paces, at one point, all the surface of the water among the stubs of the sweet flags is winking with the rise and descent of a myriad fingerlings. For hibernating rodent and hidden turtle, what dreams, I wonder, come on such a day of spring in January?

JANUARY • 29

MAPLE ICICLES A cold snap in the night ended the days of thaw. Down the street, boys are sucking icicles formed of maple sap where twigs were broken. Although the trees are Norway maples, they say the frozen sap is pleasantly sweet.

JANUARY • 30

WEATHER-WISE PINE CONE In the ups and downs of recent weather, we have been interested in watching a pine cone. It is a large longleaf pine cone that Nellie picked up somewhere in the Carolinas when we were making our trip north with the spring. We have hung it from the limb of a cedar tree near the kitchen window. In the openings of the cone we place peanut butter for the chickadees.

Watching it day after day, we have noticed how it forms a natural indicator of humidity. On dry and sunny days, the cone is open. During the recent thaw, when the air was moist with melting snow, the cone was tightly closed. Its fibers respond to the rise and fall of humidity as surely as do the leaves of the rhododendron to changes in the winter temperature. On intensely cold days, we notice how the star of pointed leaves at the end of each rhododendron twig folds down until the tips almost touch. Then, as the mercury rises, the leaves swing upward again.

Pine cone and rhododendron leaf—they are weatherwise. So tightly shut was the cone, three or four days ago, that it resisted all the paw-tugging of a gray squirrel that was tantalized by the smell of the peanut butter locked within. Finally the outraged squirrel fell to with its gnawing teeth. There is now a gaping hole, like the mouth of a tunnel, in the side of the pine cone.

JANUARY • 31

STARS OVER THE TREE About nine o'clock on this cold, still night, I pause beneath a dying maple tree. Overhead, the bare branches stand black against the cold glitter of the sky and the whole tree seems laden with stars.

CHAPTER TWO

February

FEBRUARY • 1

RETURN OF THE STARLINGS Toward dusk, on this day of wild, tumultuous wind, I follow the swamp path to the old orchard hillside where for more than a decade I have planted my Insect Garden in the spring. I walk with my hat brim down and my head tilted toward the northwest, into the wind. Branches flop together like the slapping of ropes on a vessel. One maple branch has rubbed against a wild cherry branch until both are flat on one side and as smooth as though pumiced. All over the slope, under the fruit trees, the fallen apples, decayed and soggy, are disappearing altogether, coming to their natural and fitting end, gently lowering to the soil the seeds they contain. Decay, rightly understood, can be as beautiful as growth.

It is while I am leaning against one of the swampside trees, listening to the sounds of the wind—the wind hissing through the dry phragmites, the wind booming in the maples—that I become aware of the return of the starlings. Half a hundred come in on a great gust, flying low, skimming up and over the wild-cherry tangle as a single bird, as a rider taking a hazard, then descending close to the pounding waves of the cattails. They turn and rise into the wind and other starlings arrive in little groups and flocks to join them. Minute by minute, their numbers grow. The flock swells until half a thousand, then a thousand and more, swirl and swoop over the swamp in the great wind.

Each time the weaving, drifting, plunging cloud of birds descends and then zooms up again, I can hear—clear across the swamp and in spite of the great wind blowing away from me—the sound of the thousand wings, a silken sound at times, again a murmur like wind in the branches of a pine. Up again and again the flock circles—in funnels and waves—sometimes stretching like a long, loosely constructed serpent, writhing up and down over half the length of the marsh. At other times, they sweep like a rain cloud, dragging low, with the moving curtain of the lowest birds like a fringe of falling rain beneath.

A few at a time, they drop into the phragmites where they will roost for the night. Each alighting bird clings to the stem of its wildly gyrating support. Once, when the birds are all down and the upper part of the phragmite stand is sown with the black dots of their bodies, they all leap into the sky again for a final whirl in the airy surf pounding over the lowland. I come home through the gathering dusk exhilarated as though I, too, had taken part in these synchronized evolutions in the windy sky.

FEBRUARY • 2

CONSERVATION The long fight to save wild beauty represents democracy at its best. It requires citizens to practice the hardest of virtues—self-restraint. Why cannot I take as many trout as I want from a stream? Why cannot I bring home from the woods a rare wildflower? Because if I do, everybody in this democracy should be able to do the same. My act will be multiplied endlessly. To provide protection for wildlife and wild beauty, everyone has to deny himself proportionately. Special privilege and conservation are ever at odds.

FEBRUARY • 3

HAWK AND SPARROWS During a heavy fall of snow this morning, a Cooper's hawk alighted on the top of the weeping mulberry tree where, a few weeks ago, a starling took refuge from a sparrowhawk. It hunched up motionless in the falling snow. Soon a flock of a dozen or fifteen English sparrows darted into the tree, flitting about amid the maze of drooping branches, often no more than a foot or two below the perching hawk. It sat there, immobile, with the chirping, active flock of little birds just below it but beyond its reach. Once it lifted one foot, opened and closed the talons, then became motionless again. This continued for several minutes and all the while the sparrows, safe from their great enemy, continued to chirp and dart about. Then the hawk took wing and scudded away into the curtaining snow.

FEBRUARY • 4

WORK We say we have to work so hard in order to get so little in life. That little may be more than we need. Reduce the complexity of life by eliminating the needless wants of life, and the labors of life reduce themselves.

FEBRUARY • 5

RING AROUND THE MOON Needles of ice, billions of tiny crystals and prisms floating in the deathly cold at the top of our sky, are producing a show in the heavens tonight.

Dusk has turned into early dark and all the sea meadow stretches away in one illimitable plain of blackness. Overhead, as I walk toward home, the stars are wan. And the moon shines palely within a wide glowing circle, a halo that stands out against the sky.

To weather-wise farmers, such a ring around the moon is a storm signal, an augury of rainfall at times when crops are growing. To the naturalist, it has another interest. It provides us with a glimpse into a forbidden world, a world where human life is impossible, a world of unbearable cold and rarefied air, in the world of the upper sky.

So far as our atmosphere is concerned, we, together with all animal life, live at the bottom of an ocean. We need the pressure and the radiated warmth and the oxygen of the Earth's surface in order to exist. Only in recent years have we penetrated briefly into the upper realms of the air with the aid of oxygen tanks, pressure cabins, stratosphere balloons and rockets.

It is in the thin air, miles above the earth, that our loftiest clouds are born. They are the delicate, feathery cirrus clouds that stretch in long lines and wide curves and faint streaks across the sky. Sometimes a thin white sheet extends through the upper air, an ice haze in the high sky. The lower clouds, the thunderheads, and the fog masses of the rain clouds, are formed of water droplets. But far above them, where the cirrus cloud has its home, no water can remain for an instant unfrozen. The component parts of these delicate clouds are crystals of ice.

It is the moon's rays shining on these infinitely multiplied ice needles that creates the glowing halo I see in the sky above the sea meadow tonight. Ice haze in the heavens is essential to the moon's halo. Both the ring around the moon and the halo around the sun have the same explanation. It is to invisible ice crystals, infinite in number and all reflecting light rays, that we owe the arresting sight of these glowing circles in the sky.

FEBRUARY • 6

SKUNK CABBAGE In a dozen places along the swamp trail today, I see skunk cabbage already spearing upward out of the black, partially frozen soil. Each year, its mottled spathes rise as the advance guard of spring. Within these hoods, the fleshy flowers soon will form. First the flower, then the leaf—that is the odd reversal of events in the life of this plant pioneer.

FEBRUARY • 7

SONG OF THE NUTHATCH Among the upper branches of the dying maple, about eight o'clock this morning, I heard a small bird voice. It sounded a little like the faraway, repeated "wicky-wicky-wicky" of a flicker; even more like a miniature tufted titmouse giving an abbreviated version of its call. Instead of the full "wheedle-wheedle-wheedle" of that bird, this call was a kind of "whit, whit, whit, whit," smaller, softer than the call of titmouse or cardinal. It was just loud enough to carry across the yard. The bird was a white-breasted nuthatch. I could see its slender bill opening as it called. In the woods, in early spring, I had heard the same call without being entirely sure of the singer. In quality, it is far different from the nasal "yank" usually associated with nuthatches. On this sunny February day, the male repeated endlessly, in advance of the season, his clear, mellow, diminutive song of the spring. Like the spathe of the skunk cabbage, the song of the nuthatch is an assurance that winter is nearing its end.

FEBRUARY • 8

FROZEN LIFE I looked at the woolly bears again this afternoon, each a bristly little doughnut on its bed of moss beneath the shelter of the log. Animation is suspended, life is almost extinguished, log and caterpillar seem equally unalive. For the cold-blooded, the northern winter must always be a time of deathlike sleep. In the laboratory, butterfly larvae can be frozen so hard they shatter like glass if dropped on the floor. Yet, if they are thawed out very slowly, they show no ill effects. In fact, a scientist friend tells me, they can be frozen and thawed repeatedly and still return from apparent death to certain life each time. The trick lies in the gradual change. Frogs have shown that they can stand water that is so hot it burns the skin, if the fluid is heated slowly enough. It is the sudden change, the swift alteration of conditions, that proves disastrous for individual and species alike, the change that does not allow time for adjustment or evolution.

FEBRUARY • 9

A MYRTLE WARBLER DRINKS MAPLE SAP Each winter, the gray squirrels nip twigs on our maple trees and obtain liquid from the sap that oozes from the opening. One such squirrel-produced spigot has been dripping on the driveway for several days. This morning, a myrtle warbler alighted on the twig and drank the drops of sap as they collected below it. Here was another source of winter water.

FEBRUARY • 10

POPPING OF THE WISTERIA About five o'clock this afternoon, the popping of the wisteria began. First there was an explosion like a cap pistol and then something like a small pebble hit against the window. Another shot and we heard a little thump against the side of the house. The annual barrage of the wisteria seeds had commenced.

The vine from which they come is about twenty feet from the side of the house, trimmed back to form a mound hardly more than three feet high. The seed pods are about half a foot long, brown, stiff, velvety on the outside and holding from four to eight seeds. Each seed is roughly round, about half an inch in diameter, flat on one side and curved slightly, like an airplane wing, on the other. They are hurled for surprising distances like rotating wings or miniature discuses.

I touched one pod and it popped as though I had pulled the trigger of a gun, hurling the seeds away. Apparently the two halves of the pod are under tension so they split with a twisting motion. The following day, I noticed that the pods had all twisted like screws where they had fallen to the ground. The seeds, skimming in the manner of flat stones, sailed for upward of fifty feet. One struck among dry leaves with a rattle that reported exactly the point where it had landed. Another struck me on the cheek with a stinging blow when I was more than fifteen feet from the vine. The torque, or twisting tension, of the pods, probably supplies the driving force for the flying seeds.

This wisteria cannonading continued from about five until eight, long after dark. Why today? Why this particular time of day? The temperature was about 42 degrees F. at the start and about 40 degrees at the end. In all probability, the trigger that nature pulled was some finely balanced combination of age, humidity and temperature. About two-thirds of the pods on the vine split and fell during those three hours. By bedtime, we found the ground beneath the vine littered with the "empty shells" from the cannonading, the seed pods, the halves separated and twisted.

At times, during the popping of the wisteria, we would hear two shots almost simultaneously. Sometimes we could hear the seeds scaling close past us through the air. All over the yard we found the flat, brown seeds scattered. Each was about the size of a nickel. If these seeds, scaling away from a low bush, traveled half a hundred feet or more, how far would seeds from a high wisteria, climbing up a porch or tree, have gone?

FEBRUARY · 11

SCATTERED SEEDS Driving around this morning, I noted other wisteria vines, some a mile away, with the ground below littered with shattered pods. Everywhere in this region, the vines had responded in the same way at the same time to the same combination of circumstances.

FEBRUARY · 12

OVERWINTERING TREE SWALLOWS Near Captree, on the south shore of Long Island, today, I came upon a flock of more than fifty overwintering tree swallows. They were resting on the concrete of the parkway, creating, with their massed bodies, a kind of avian sand bar extending out from the north side of the road. The sunshine was bright upon them. In the blue of each back there was the glinting of metallic green. They rose in a compact flock and turned with the pure white of their under plumage shining in the sun. There was plentiful food all around on the bayberry bushes, gray with fruit, and beyond rose dense stands of plumed phragmites offering shelter for the winter nights.

FEBRUARY • 13

LENGTH OF LIVING How strangely inaccurate it is to measure length of living by length of life! The space between your birth and death is often far from a true measure of your days of *living*.

FEBRUARY • 14

TIME SENSE IN A NUTHATCH During the early part of this winter, we had but a single suet-holder hanging in the yard for woodpeckers and nuthatches. Starlings soon learned to alight on the swinging feeder. They dominated it, driving away the white-breasted nuthatch day after day. The bird soon learned to come just at a certain time of day when the starlings were gone. Only then we saw it. Now we have more than one feeder and the nuthatch is able to get suet at one, while the starlings are feeding at another. It appears now several times a day.

FEBRUARY • 15

HOW LONG IS A BIRD'S MEMORY? The coming of the nuthatch to our suet-holders has recalled an experience John Kieran once related to me about a white-breasted nuthatch in his yard.

On the limbs of three trees back of his home in Riverdale, he was in the habit of hanging three suet-feeders. Each year, he took the feeders down for the summer, putting them up again late in autumn. One year, a few days before the feeders were in place, he saw a nuthatch fly into the yard. It darted directly to the limb where one of the suet-holders had hung the winter before. Then it flew on a beeline to the position of a second feeder and on to the place where the third feeder had been located.

There was nothing to attract it to those particular limbs except memory. The bird had been there before. Its action identified it as a previous visitor to the yard as surely as though it had been banded. It had remembered, at least since the winter before, exactly where the feeders had been. How long is a bird's memory? The answer undoubtedly varies with different species. But for the white-breasted nuthatch, it extends across summer without question from winter's end to autumn.

FEBRUARY • 16

LOVE SONGS The time of tomcats is at hand. I heard my first serenade last night. Earlier than the first robins, earlier than the first violet, often while snow still lies on the fields, there is this familiar vocal harbinger of spring, the yowling of the tomcats.

Love comes early to the breast of the feline. Its back-fence love song, the fire that stirs it to battle, is only the beginning. As spring advances, creatures of wood and field and pond, each in its own way, will join in a swelling chorus of serenades. Only a few days now, and I will walk abroad some morning and hear the exuberant rat-a-tat of the flicker's bill on hollow limb or seasoned telephone pole or even on the metal of a street-lamp reflector. This triphammer solo is the love song of the flicker.

A few years ago, one of these woodpeckers found a hollow metal cap at the peak of the reptile house at the Bronx Zoo, in New York City. Day after day, its courtship tattoo filled the building below like the reverberations of a pneumatic hammer. Keepers tried to shoo the bird away. It simply hopped over the ridge of the roof and began drumming on the other side of the metal cap. This kept up for weeks and ended only when the ardors of courtship abated.

The wild serenades of the spring will assume innumerable forms. Wildcats and lynxes, like their backyard relatives, caterwaul in the night. Even the solitary porcupine is overcome by tender emotions and wanders through the woods singing a little love song, a quavering serenade that resembles the rising and falling wail of an infant.

Of all the varied bird songs that the season will bring forth one of the most delightful and little-heard will be the spring song of the woodcock. It comes at evening, just at dusk, at a time when the last melodies of the diurnal songbirds are dying on the soft air. Then, from some open field the shy woodcock begins whirling aloft to tumble down out of the darkening sky uttering the sweet twittering notes that comprise its mating song. This is its only music of the year.

FEBRUARY · 17

WINTER BUDS Walking in the bare woods beside Milburn Pond, today, I find myself nibbling winter buds like a feeding grouse. Except for sassafras, most of the buds I sample are remarkably tasteless provender. A grouse must be content with nourishment alone.

FEBRUARY · 18

THE SHORN LAMB How many perversions of observable truth have been cherished in the sayings of past generations! "The wind is tempered to the shorn lamb." "God builds the nest of the blind bird." The wish is father of the thought; the wishful thought becomes the epigram. Yet all the while, resting on an untruth, it is, as another saying has it, a snug nest on a rotten limb. It is morally as bad not to care whether a thing is true or not, so long as it makes you feel good, as it is not to care how you got your money so long as you have got it.

FEBRUARY · 19

CAT FLEAS IN A CAT FIGHT The copious yowlings of a cat fight awakened me about four o'clock this morning. As I lay there sleepless, I recalled Samuel Butler's notes for *Erewhon Revisited*, his return trip to the imaginary land of Nowhere spelled backwards. There, Butler recorded, educational institutions employed professors of all the languages of the principal beasts and birds. His host was Professor of Feline Languages. At the time, he was teaching The Art of Polite Conversation among cats.

As threat and counter-threat rose to a crescendo outside my window, I reflected that this was not Erewhon. I remembered a book I had seen listed in the New York Public Library. The card in the file carried the title: *Pussy and Her Language*. I must get out that book sometime. With sound and fury, the two back-fence warriors came to grips. It occurred to me that a cat fight must be a time of uncertainty and peril for cat fleas. How many a flea loses its life unseen and unsung during such a battle! It occurred to me that right now, outside the window, fleas were living dangerously amid the flying feet and raking claws and biting teeth and lashing tails and rolling bodies. It further occurred to me that I probably was the only human being among all the billions living on this planet who was at this moment, or ever, concerned with the fate of cat fleas in a cat fight. I am saddened at times when considering the smallness of things with which my mind is occupied. As Dr. Leland O. Howard once said to me: "People think entomologists have small minds because they interest themselves in small animals." I turned over and, in time, went to sleep, leaving the fleas to their fate and the cats still yowling.

FEBRUARY • 20

THE BIRD OF SPRING It was not cats in combat that awoke me this morning. Instead it was a glad tumult of sound in the tree tops. More than a hundred redwings had come back at dawn. Every one was a male with scarlet epaulets flashing. Their liquid calls, those "Okaleees" of spring, so old, so ever new, lift our spirits beneath the February sky.

To some, spring is symbolized by the return of the robins, to others by the earliest green of grass pushing up in the wake of the snows, to others by the long, tremulous calling of the marshland peepers. But to me the return of another spring is marked by the coming of the redwings. Each year, usually a little after the middle of February, the male blackbirds come back with a rush. Almost overnight, the dead vegetation of the winter swamp is filled with life. For all who live within sound of a marsh, the redwing is the bird of spring.

Later this morning, I followed the swamp path. Already, above the yellow flags and cattails, the redwings were swirling like windblown leaves. When they alighted, riding on swaying stems, they were still half in the air. And when they took off from cattail heads, the fanning of their wings sometimes formed below them an insubstantial cloud of flying fluff. The air rang with their wild xylophone calling. Everywhere the males were singing, darting, chasing each other, shooting up like rockets, whirling like pinwheels.

The great work of these early days is finding and defending a territory. The males come north first. They are the advance guard, the pioneer settlers. By the time the females arrive, weeks later, they have divided up the land. Each stakes out a homestead, so to speak. Then with swift dashes and spectacular aerobatics, it defends and holds as much of its chosen area as possible. In this manner, the nesting blackbirds are spread out over the length and breadth of the swamp.

As I watched, one male that had just vanquished a rival swooped to a landing a dozen yards away. Its expanded epaulets were shining and its rolling call seemed like an exultant "OK—Meee!" Henry Thoreau thought the redwings of the Concord River cried "Conquereee!" Others have suggested that the call sounds like "Gurgleee!" At any rate, it is a call that is exultant and jubilant and filled with vitality.

Heard over the drear stretches of dead vegetation that extend across the marshland on this February day, it is a sound that lifts the heart of the hearer. It carries like a bugle call across the battlefield of winter's defeat. The exulting song of the redwing is a fitting voice for a season of flowing sap and awakening life.

FEBRUARY • 21

FAUNAL DESERTS I have been remembering a beautiful woodland I enjoyed for years, now wiped out forever. A few men, a forest; many men, a desert. That old saying never had greater point than it has today. It remains to be seen whether Man is ultimately known as the Lord of Creation or the Lord of Destruction. Given high enough population and the destruction of man may balance the production of nature. It is only our idea that man was put on earth to subdue the other creatures inhabiting the planet; to kill off one, then another, until he has produced a faunal desert around him. He has the privilege of living on earth with a host of other interesting forms of life. Enjoying that privilege is part of the rich and rounded existence he is permitted to know.

FEBRUARY · 22

THE EARLY ANT Abnormally warm today, the thermometer in the high fifties. Grayish gnats are hovering in the sunshine at the Insect Garden. Looking at the hanging suet-feeder in the maple in our backyard, I find that small brown ants have climbed the tree and are dining on the suet. These suet-eaters are the first ants I have encountered—the earliest ants of the year.

FEBRUARY · 23

CLOUD FEATHERS About eleven o'clock this morning, on looking up at the sky, I noticed a curious formation of high cirrus clouds such as I had never seen before. Extending across a great part of the heavens, from southwest to northeast, stretched two slender lines of white. Across them, feathery clouds ran at right angles. The effect produced was that of two gigantic white feathers, with central shafts and plumy barbs, floating side by side in the sky. Their substance is crystalline ice, frozen vapor such as, not long ago, formed the ring around the moon. In all probability, the crystals had been distributed into their feather forms by the great winds that sweep across the upper reaches of the sky.

FEBRUARY · 24

REDWING MORNINGS The redwings were in the trees again at dawn. Each morning they come to feed on grain we scatter in the yard. We awake to their liquid calling, a song well suited to a time of unlocking ice and brimming streams. As the crow, cawing over snow-covered cornfields, is the bird voice of winter, so the redwing, exultantly calling over the thawing marshlands, is the bird voice of the earliest spring.

Under our maple trees, the blackbirds feed as a flock. But when they fly away toward the swamp, it is the beginning of a kind of Oklahoma land rush. Each male is intent on staking out and defending a territory, intent on being a landed proprietor when the females arrive later on. By getting this preliminary division of territory completed in advance of the return of the females, valuable time is saved. The days are none too many for mating, nesting, raising broods, getting the fledglings sufficiently grown for the long autumn migration. This arrangement—first the coming of the males, then the division of territory, then the arrival of the females, then the nesting time—the unvarying sequence of these events is an essential feature of each yearly cycle of redwing life.

FEBRUARY · 25

A PLUNDERED COCOON Where the swamp path runs past a clump of arrowwood, this afternoon I found a cecropia cocoon with a gaping hole in its side. The pupa within was gone. Who had dined on it—rodent or bird, mouse or woodpecker? The position of the hole provides the answer. When a mouse invades a cecropia cocoon, it gnaws a hole near the bottom. When a downy woodpecker attacks a cocoon, it uses its chisel bill to cut a hole in the side. The cocoons that are safest from the aerial raids of these woodpeckers are those at the tips of the branches. Under the bill strokes, they swing like punching bags, thus blunting the power of the blows and often saving the pupae within.

FEBRUARY · 26

CAMOUFLAGED MEADOW LARKS Three meadow larks flew into the yard this morning. Several have overwintered on the wide sea meadows south of us. As the birds alight, and the sudden flash of their white tail feathers disappears, I am struck by the perfection of the camouflage provided by their color and pattern of plumage when seen from above. As I look down on them from an upstairs window, they seem, except when they move, part of their surroundings. We usually think of meadow larks, with their striking black and yellow breasts, as colorful birds. And so they are from a low viewpoint. But from the higher viewpoint of the soaring hawk, they are the reverse. And, on the wide sweep of an open meadow, it is from above that an attack is most likely to come. Meadow lark camouflage is partial camouflage, camouflage where it is needed most.

FEBRUARY · 27

SNAKE DREAM A man who had hunted bushmasters in the Panama jungle visited me this evening. He recalled an experience that throws some light on the inception of dreams.

Late one night, in the jungle, he awoke dripping with sweat, tangled up in his mosquito bar, one hand bruised and bloody. He remembered vaguely a terrible nightmare in which he was struggling with a snake. Less than a week later, he had the same nightmare. This time he awoke pounding one hand on the ground with his other hand. He puzzled over what was causing the dream. Awake, he had no fear of snakes. He had cornered and captured bushmasters, coral snakes and other deadly serpents for years and had slept untroubled. With natives, he had even run barefoot through bushmaster country at night hunting wild pigs. Now, he dreaded to go to sleep for fear the nightmare would return.

A doctor he consulted could offer no suggestion for overcoming the nightmares except to keep out of the jungle. Then, one night, he awoke just as the nightmare was beginning. He discovered that, in going to sleep while lying on his side, he had taken hold of one of his wrists with the other hand. As he dozed off, the wrist within his grasp, feeling like the neck of a held snake, had set off his dream. Struggling to kill the nightmare serpent, he had battered his hand on the ground. After he had thus discovered the cause of his dream, the nightmare disappeared and never came again.

FEBRUARY · 28

REDWINGS IN THE SNOW The temperature dropped swiftly in the night and this final day of February dawned under leaden skies. Snowflakes began sifting down, at first descending negligently, tentatively, widely dispersed. But soon the drift from the sky was veiling the trees. As I walked across the yard through this chill, swirling veil of white, I heard redwings in a maple overhead. Their forms were dim and gray amid vague branches seen through the falling snow. But their cheery call came down distinct. The rolling "Okaleee!" of spring, it resounded an undaunted promise in the midst of storm.

CHAPTER THREE

March

MARCH • 1

FOOD TRAILS A small red house ant climbed a leg of the gas range this morning. On the porcelain top of the oven, it discovered a drop of bacon grease. At full speed, it turned, ran back down the leg and disappeared. Within two minutes, a line of a dozen red ants was running up the leg. Like bloodhounds on a trail, they followed the footsteps of the scout directly to the discovered food. It has been proved, by marking ants, that the scout does not lead such a party. The other ants go unerringly to the spot themselves. How?

Only very recently has the true explanation been known. Scientists have discovered that when an ant finds food it becomes excited and this excitement causes anal glands to give off secretions that lay down a special kind of trail that other ants can follow through their sense of smell. Such "food trails" are entirely different from the ordinary formic acid trails that many ants lay down as guiding roadways.

They have a special meaning for every ant that comes upon them; they provide an irresistible invitation to follow the food trail and join in the harvest.

MARCH • 2

CROW SHADOWS Bright skies above the fields of snow. Moderating weather has melted portions of the snow on the hillside above the swamp. This afternoon, three crows fly slowly over the hill and into the wind. It is interesting to watch their shadows trail across the patches of snow, then disappear on the dark bare ground, then reappear on the snow beyond the open space.

KATYDID WINGS An artist on Cape Cod has written me a letter about katydid wings. In *Grassroot Jungles*, I mentioned an experiment by U.S. Department of Agriculture men that led them to conclude that a katydid rasps its wings together between 30,000,000 and 50,000,000 times during the summer season. Why, my correspondent inquires, are not the wings completely worn out before the summer is over? No machine could possibly stand up under such wear without lubrication. What explains the ability of katydid wings to endure such constant and long-continued friction? The answer lies in the amazing properties of chitin. Light and strong, resistant and enduring, this material forms the outer covering of every insect on earth. It provides strength to the armor plate of the beetle, keenness to the lancet of the mosquito, endurance to the rasping fiddle and bow of the cricket and the katydid.

MARCH • 3

SQUIRRELS AND BLUEJAYS In the rose garden today, a squirrel dug a hole among the fallen leaves, dropped in a peanut within its shell, covered in the hole, patted down the earth and hopped away. He had gone only a dozen yards when a bluejay dropped from a branch overhead, alighted at the exact spot and began tossing aside leaves with its bill. It hunted for a long time but it never found the buried nut. Apparently, the squirrels outwit the jays by digging holes too deep for them to uncover. I notice that when I throw out peanuts, the squirrel comes running when it sees a bluejay descend to the grass and the bluejay comes flying when it sees a squirrel make a dash for nuts. The squirrel watches the bluejay and the bluejay watches the squirrel and they both watch me!

MARCH • 4

I TURN A KEY When I turn a key, I have but the vaguest idea of what happens inside the lock. When I step on the starter of my car, I am a bit bewildered by the thought of all the things that begin turning in consequence. When I twist the knob of my radio, the results are produced in a manner that is still baffling to me. I am part of an age of invention, an era of great mechanical progress. But, like millions of my fellowmen, my part is largely parasitic. I move with the ship but it is somewhat in the capacity of a barnacle. Only as a consumer, a buyer of inventions, can I claim any contribution to technological advance. Born no doubt of ignorance and the feeling of insufficiency, vague misgivings fill my mind when I am in the midst of machines. It is when I am among the old, rooted, growing, living things of nature that I feel most at home.

MARCH • 5

THE ROBIN'S RED Robins are everywhere this morning. They are running over the wet March ground, in yards, among the open fields, on the swamp edge hillside. They are darting from limb to limb in the old apple orchard. Like scouts or an advance guard, a few red-breasts have come up from the south during recent days. But now the glad robin invasion is in full swing.

With bills of brightest yellow and breasts of shining red, the males are in their full breeding plumage. How red is the red of the redbreast? Exactly what color is it? The answer seems simple. Yet, as a writer in

Fauna Magazine pointed out some years ago, authorities are far from agreement.

In his *Handbook of Birds of Eastern North America*, Frank M. Chapman says the robin's breast is "rufous." Elliott Coues, in his *Key to North American Birds*, calls it "chestnut." Edward H. Forbush, in *Birds of Massachusetts and Other New England States*, simply refers to it as "red." J. Fletcher Street puts it down as "rufous" in his *Brief Bird Biographies*. Montague Chamberlain, in his *Handbook of Ornithology*, calls it "brownish red" while, in *Birds of America*, T. Gilbert Pearson gives it as "plain, deep cinnamon."

Years ago, in the General Electric Laboratories, at Schenectady, New York, I was shown a photoelectric colorimeter, a complicated device for enabling printers to match colors exactly. With such an apparatus, the question of the exact color of the robin's red could be settled. Or could it? I suspect that different robins vary in hue more than our eyes detect. But in their breeding plumage, there must be a common denominator that can rightly be called the red of the robin.

With such a machine at the disposal of a naturalist, how many other things of interest and exactitude could be determined! The leaves of the same tree—how widely does their green vary? And "grass green"—how infinitely varied is it? How many hues are there in the wing of a polyphemus moth? Such questions as these, with the aid of a photo-electric colorimeter, could be answered precisely for the first time in all the years man has been gazing with delight and fascination on the world of color around him.

MARCH · 6

SQUIRRELS GATHERING CEDAR BARK Outside my study window, a gray squirrel is stripping off the dry bark of cedar limbs and carrying the shreds and pieces away. This is an activity I see each year. Apparently the bark goes into nest building as the time of baby squirrels is at hand. Is the squirrel outside my window carrying away cedar bark because it is dry and without pitch, or because there is in the little rodent brain of the gatherer some untaught wisdom that leads it to this particular tree? Cedar chests protect clothes from moths. The odor of cedar, which seems pleasant to us, is repellent to them. Would cedar bark, added to the nest of the squirrel, provide a natural insect repellent that would help keep it free from irritating and injurious parasites?

MARCH · 7

PUSSY WILLOWS IN THE SNOW I come upon a pussy willow bush this morning, where all the silky, silvery aments are decorated with immaculate white. Granular flakes, as fine as confectionery sugar, sifted from the sky during a little flurry in the night. Each downy catkin wears a shining cap of snow.

MARCH · 8

ARCHANGELS AND ANTS I read in an evening paper today a letter from a reader announcing, to his own satisfaction, a great discovery. All the animals on earth, he says, are directed by archangels, each group having a special archangel over it, controlling its activity. The Archangel of the Ants, for example, imparts its wisdom to these small creatures. As a result, the writer observes, they are so intelligent in their actions that they resemble humans, "keeping slaves, getting drunk and conducting wars." That such refinements of conduct should be placed at the door of an angel of any kind, let alone an archangel, apparently offers no difficulty to the author. Nor is any proof deemed necessary to bolster up the assumption that rings so soundly in his ears. The assumption is sufficient. Let the world *disprove* it!

However, the seeker after truth has a harder and longer road to travel. In all sincerity, he must rather be right than President. Like J. Henri Fabre, the grand old insect-explorer of France, he must be a Doubting Thomas who demands proof over and over again. In every search for the truth in natural history, the invariable rule should be to seek the simple explanation first. Only after the simplest of theories has been disproved should the farfetched hypothesis be considered.

MARCH · 9

REDWING FIRE At dawn to the swamp to see the redwings. Neither the overcast sky nor the chill March wind can subdue their exuberance. They spin and tumble in aerial jousts all over the gray-yellow prairie of the cattails. The red of their epaulets shines like coals. Redwing fire burns brightly this morning all across the swamp. In the early light, one blackbird, at the tip of a cattail, fluffs himself up until he resembles a crow. I have to look twice to be sure. These are the expansive days of the redwing year. Here is one bird that looks almost as big as he feels!

MARCH · 10

THE REDOUBTABLE RODENT Never underestimate the resources of a squirrel. This is the chastened advice of one who did. The other day, a friend of mine told me of a relative who had fed nuts to a squirrel that came each morning to the window by his breakfast table. It learned to take the husk of candy off a Jordan almond and it welcomed cookies that contained nuts. But one morning, when its benefactor offered it a cookie with no nuts in it, it bit him. Outraged by this ingratitude, he got revenge by propping a nut-filled cookie up against the inside of the window where, for several days, the disconsolate squirrel could look in and see it but could not reach it.

I remembered that story this afternoon when the redoubtable Chippy, most fearless of the squirrels in the neighborhood, came to the kitchen window. After feeding her a peanut or two, I teased her by closing the window and placing a nut inside the pane. Somehow, the animals I play practical jokes on never seem to have a sense of humor. Instead of peering in dolefully, Chippy attacked the outside of the window frame like a buzzsaw. Before I could get the window open, chips and splinters were flying in all directions and gouged-out places had to be puttied and painted in the course of time. The next time I play a joke, it won't be on a squirrel.

Perhaps I should have been forewarned by an event of the winter. A stray tomcat wandered into the yard one morning and promptly chased Chippy into the neighboring wisteria bush. But she did not stay in the wisteria bush. Before the cat could move, she leaped onto his back, nipped him hard, leaped back into the bush again. It was the fastest movement I have ever seen a squirrel make and it sent the amazed cat fleeing for his life. A redoubtable rodent is Chippy!

MARCH • 11

PUSSY WILLOWS IN THE RAIN A fine drizzle of rain descends from a dull gray sky through the windless air most of the morning. The pussy willows, which I saw not long ago in snow, seem to shine, to catch all the light around them, their silver fur adorned with glistening droplets. As many as half a dozen drops cling to a single catkin. Each droplet is crystal clear and a number, apparently flattened where they touch the aments and rounded on the outside, act as little magnifying glasses, enlarging the silver-gray hairs beneath them. One catkin is encircled with a series of shining drops as with a string of transparent pearls. As I watch, from time to time droplets join to form larger drops. The lower part of each larger drop glows with a kind of captured and concentrated light, as though the richness of the illumination were collecting at the bottom of the pendant liquid. I count the aments on the bush. Slightly more than six feet high, with but seven main branches, the little tree holds aloft 859 catkins, each shining and spangled with droplets of rain.

MARCH • 12

YELLOW BILLS The yellow of the marsh marigold is no more a sign of spring than the yellow of the bills of the starlings. Dark gray in the winter, they turn butter yellow as the breeding season approaches.

But how does a bill change color? Does it become yellow all the way through or just on the outside? Is a new bill grown? Does it change its color all at once or does it begin to get yellow at one end first? If so, which end? The base or the tip?

The answer seems to be that the starling's bill begins to change color at the base and the yellow spreads outward to the tip. A bill, incidentally, is not solid as though carved from some plastic substance. It has layers. The one on the outside is transparent. It is in the layer beneath this layer that the pigment is added and the change in color takes place.

MARCH · 13

THE BEAUTIFUL TREES One of the most suitable places in the world for lonely reflection is in a packed subway car. How alone you are in the midst of a crowd of strangers! In the city today, I look at the passengers around me, all lost in the details of the latest holdup and homicide. I fall to imagining the acres of forest, sun-flecked and wind-blown, that are laid waste to provide New York with newspapers for a single day. How many beautiful trees gave their lives that today's scandal should, without delay, reach a million readers!

MARCH · 14

SWAMP DUSK Under a sunset and clearing sky, after a day of rain, Nellie and I watch the birds come home to roost in the phragmites and cattail stands of the marsh. Here redwings, starlings and grackles spend the night. The blackbirds arrive with a flourish. They dive and twist in superb airmanship as they descend. The starlings make more direct, businesslike flights devoid of show when they come in singly or in small groups. And the purple grackles labor along, seeming to drag their great tails through the sky with difficulty. The flocks increase and the clamor grows. It may be that, in the deepening dusk, the vast noise of the birds that have already arrived may act as a sound signal, directing stragglers to the site of the roosting place.

Listening there, on the edge of the swamp, we discover there is, in the discordant tumult of the hundreds of birds, a noticeable beat or rhythm with a high recurring note. At times, we seem to hear another bird's voice—a sound lifting above the tumult of smaller sounds like the faraway, confused calling of Canada geese. It becomes so distinct at times that we search the darkening sky for V's of flying birds. But none is there. This is the only time we have ever heard anything of the kind. Perhaps we are fortunate in hearing a combination, the right combination, of blackbirds, starlings and grackles, all calling at once and producing the recurring overtone or beat that carries to our ears as the crying of distant geese.

The clamor grows less as the dusk deepens until, in the end, the softer trills of the blackbirds and the occasional shorter calling of the other birds seem a kind of avian lullaby. Perched among the cattails and phragmites, the birds sound as though they are singing themselves to sleep, their feet locked in position for the night on the upright stems.

MARCH · 15

PUSSY WILLOWS IN THE MIST A mild, misty morning, the air silvered with fog. Everywhere the green of new grass. The door of spring has swung more widely open in the night. I stand once more beside the pussy willow bush. I have seen the catkins powdered with snow, spangled with raindrops. Now I view them surrounded by mist. Although I observe no droplets on them, they appear saturated, swollen. The fine moisture of the mist fills them. Viewed against the east, they glow in the glowing light, seem larger than before.

MARCH · 16

A GRACKLE DUNKS A RAISIN Each day, now, we see in our backyard an interesting instance of a habit entirely hereditary. Blackbirds, starlings, cowbirds and purple grackles alight to feed on scattered grain and pieces of bread. Each time a grackle picks up a crust in its bill, it walks sedately to the bath and dips the food in the water before swallowing it. This morning, among the food was a slice of raisin bread. One grackle picked out one of the raisins. It carried it half across the yard and dunked it carefully before it swallowed it. No other bird we see does the same. None of the neighboring starlings or redwings or cowbirds ever imitates this act or acquires the habit. Like the raccoon that washes its food, the grackle acts instinctively.

MARCH • 17

SASSAFRAS TWIGS In these days just before the official beginning of spring, the signs of the season are increasing everywhere. Walking along the edge of Milburn Pond today, I notice the sassafras buds, how soft and swollen they have become. They are filled with juice and flavor. As I follow the pond shore, I nibble at the buds and chew on the green twigs.

Among the sights and sounds and smells and tastes that have special power in stirring our memories and emotions—sights like moonlight on a misty river, sounds like church bells far away at sunset, smells like leafsmoke in the autumn dusk—among these belongs the wild flavor of sassafras.

From topmost leaf to buried root, the sassafras tree is a storehouse of flavor. What walker in the woods has not known the pleasures of chewing its spicy twigs and buds? On this day when sap is running again and spring is swelling the twig tip buds, I find a fresh delight in the old, familiar taste.

Sassafras was the first product exported by the Pilgrims to Europe. It is a spice of the New World. Even today, little bundles of reddish rootbark can be purchased in certain stores for the brewing of sassafras tea. If the potency of this beverage as a "clearer of the blood" is less highly considered now than in former times, the woodland wildness of its flavor has the same charm as of old.

In the Indiana dune country where I spent my boyhood, sassafras trees filled the fence corners and stood in little oases on the sandy pasture hillsides. I was initiated young into the joys of the sassafras twig. I have nibbled bushels of buds and browsed over acres of sassafras saplings during later years of wandering over many states. Even in midwinter, buds and twigs are aromatic and rich in flavor.

Although all the sassafras trees of my acquaintance have been of moderate size, in the South the trunks of some of these members of the laurel family rise to a height of almost 100 feet. Beneath the deeply furrowed bark, the wood is yellow-brown. Although it is soft, weak and rather brittle, it has the ability to withstand decay in soil or water. Hence it is utilized for fence posts and sometimes for boat building. In the main, however, it is as the source of the oil of sassafras—used in perfume-making and as a household remedy—that the tree has its greatest value.

A final virtue of the sassafras remains to be mentioned. For anyone like myself, who remembers the scientific names of plants and animals only after considerable application, the original binomial given to the common sassafras comes as a welcome relief. It is simply *Sassafras sassafras*.

MARCH · 18

A DISCOVERY A boy called this afternoon excited by the discovery of a rare, beautiful, exotic bird in his backyard. It shone in the sunlight with brilliant metallic sheens. It seemed iridescent. He thought it must have escaped from some zoo as he had never seen a bird like it before. The bird was a purple grackle in full breeding plumage. It is relatively common. But the boy *had* made an important discovery. He had—for the first time in his life—really *seen* a grackle.

MARCH · 19

SONG OF THE FOX SPARROW Only one day more until the beginning—the official beginning—of spring. We go to the woods north of Milburn Pond this morning to hear the fox sparrows sing. For only a few days more will they be scratching among the leaves, flashing the bright fox-red of their spring plumage among the bushes, singing that sweet, slurring little song that we hear only at the time of the spring migration. Here are no virtuoso trills, no long-held notes. But while the songs of different fox sparrows vary greatly, the slurring of the notes is characteristic. At times, we are reminded somewhat of the singing of a warbling vireo.

MARCH · 20

THE FIRST DAY OF SPRING And this is spring—officially spring! What an anticlimax! Gust-driven rain is slashing the trees under a sullen and sunless sky. The air is raw and chill; the thermometer has dropped to the low forties. Walking the mile for the morning papers, I recall Henry Van Dyke's sage observation that the first day of spring and the first spring day are not always the same thing. But under an umbrella in the downpour, I am snug and protected and isolated. I seem in a little mobile house—like a snail in its shell.

MARCH · 21

EARLIEST POLLEN The edge of the swamp stream glistens black where the wet muck shines in the sun. I stare down at tawny cattail fluff drifting away on the brown current. Just visible now are the tips of the green sword leaves of the sweet flags thrusting above the swamp earth. A honeybee crosses the stream, hovers in the sunshine, then lets itself down to sweep low over the first vegetation of spring. Its humming recedes. Then, suddenly, puzzlingly, it magnifies in volume.

Weeks ago, skunk cabbages began spearing their way upward through the black soil of the stream edge. Now they stand with fleshy flowers in full bloom within their protecting spathes. The honeybee has drifted, on its little cloud of blurring wings, into the doorway of one of these spathes. The hood forms a bandstand in miniature. Its curved interior is projecting the hum outward through the open side, concentrating and magnifying the sound.

I peer into the opening. Minute swamp flies, already attracted to the flower of the skunk cabbage, have scattered at the entrance of the bee. I see them move behind the protection of the flower. The bee hurries from pollen mass to pollen mass. Then it takes wing and its magnified hum is suddenly diminished as it emerges into the open air again.

Thus from plant to plant it goes. Sometimes, as it enters a spathe, its humming is abruptly muted. The hood of this plant faces away from me and projects the sound in another direction. In less than five minutes, the bee's pollen baskets are packed. It zooms over my head and sets a course up over the hillside to the west. For a time, I can follow the glint of its wings and body in the sun. Then it is gone. It is returning to the hive with the earliest pollen of the year, food for the first bee larvae in the cells. Each spring, it is the pioneer plant of the swamp, the skunk cabbage, that provides first harvest for the bees.

MARCH · 22

THE LOST CLOTHESLINE During the windstorm two days ago, a clothesline went down in the backyard. This morning, we saw one of the redwings hover in the air for a moment where the clothesline had been, as though searching for the missing perch, and then fly on. It was the only one, of all the birds that have alighted on the rope, that seemed to miss its presence.

MARCH · 23

BIRDS THAT FLEE THE SPRING Almost insensibly these days, the birds of the winter are slipping away. They are hardly noticed as they depart. The tempo of spring is speeding up. The shift of the seasons gains momentum day by day as the sun swings north. The green of new grass grows richer in the meadows. And soon the immeasurably varied songbirds will come trooping back from the south. In the midst of this gay renewal of life, we suddenly stop and notice that winter friends among the birds are gone. Junco and tree sparrow, chickadee and nuthatch—their numbers already have diminished or they have disappeared altogether. Already the snow buntings, with their white wing patches, and the redpolls, with their black chins and bright red caps, have followed an aerial trail to the north. The tinkling song of the tree sparrows is now largely a memory of winter.

And missing soon from the short-turf fields and open beaches will be the northern horned larks. They will leave behind them the prairie horned larks, birds that nest on the south shore of Long Island so early that the newly hatched nestlings sometimes huddle together in depressions in the ground surrounded by a late fall of snow.

The weeks that lie immediately ahead will see the swifter disappearance of the remnants of the birds that flee the spring. The present days are days of sparrow songs. At every dawn, the whitethroat sings its sweet, sad, minor melody. Each bird seems rehearsing, so to speak, for its singing at its northern breeding grounds. Only a short while more, and these singers, too, will disappear, leaving fields and woods to migrants from the south. Of all the winter birds, drifted away or soon to go, I will miss the whitethroats most.

MARCH • 24

PAPER COLLECTING BY SQUIRRELS Up the trunk of the silver maple tree and into a hole near the top went Chippy, the gray squirrel, this morning. In her mouth, she gripped a mass of newspapers, apparently destined for a nest inside. Several times, in recent days, I have seen squirrels carrying bits of newspaper. As like as not, they would take any kind of paper but, remembering the annual harvest of cedar bark for nest building, I am wondering if they have a preference for newsprint. A friend of mine always wraps clothes in newspapers before placing them in bureau drawers because she has discovered that something about them, probably the ink, is disagreeable to moths and keeps them away. Would squirrels in moth-repellent surroundings also be more free from vermin? I don't know. I do know that cedar bark and newspapers have been going into squirrel nests hereabouts this spring.

MARCH • 25

SHADOW BIRDS At noon today, a redwing hovered overhead and its shadow appeared clear-cut on the ground beside me. Without looking up, without hearing the bird call, I could have identified the shadow. How many birds can we recognize by their shadows alone? The ones that come to mind offhand are the crow, the herring gull and the turkey buzzard.

MARCH • 26

THE POWER OF GROWTH At sunset, after a day in my study, I follow the swamp path to Milburn Pond. Only a few weeks ago, when I walked this way, the wind was the winter wind, cold and desolate. The frozen earth appeared hard and dead, without hope. Yet all the power of growth, all the wealth of summer's lushness, was there in compressed and tabloid form—in seed and root and winter bud.

Now I see around me the beginning of a flood of life that nothing can halt. Seeds have expanded and split. Sprouts have driven upward toward the light. All the noiseless, resistless push of spring has now begun.

In the past, indoors and out, I have seen many evidences of the power of growth. Peas, planted in a flower pot, once lifted and thrust aside a heavy sheet of plate glass laid over the top. When thick glass bottles were filled with peas and water and tightly sealed, the germinating seeds developed pressures sufficient to shatter the glass. In slow motion, an explosion—comparable to the explosion that takes place rapidly within the cylinder of an automobile—had occurred within the bottles.

One autumn afternoon, south of Savannah, Georgia, I walked through a small bamboo grove behind a filling station. There I came upon a memorable instance of the power of growth. Thrusting upward through a pine stump, two feet or more in diameter, was a bamboo shoot. It had been forced into the stump from the ground below, driven like an iron-tipped lance through the solid wood by the pressure of growth.

Because growth in plants is usually a steady and gradual thing, we often overlook the power that is contained in the rising shoot and the expanding seed. Even when growth is not involved, the mere swelling of seeds may develop surprising pressures. Recently, a friend of mine at the American Museum of Natural History told me of a curious way in which the unhurried, increasing pressure of seed expansion is employed in the laboratory.

In some of the large animal skulls, the bones are fitted together so tightly they are almost locked in place. Forcing them apart often results in fractures. So museum workers resort to swelling peas. They pack the skulls with dried peas and place them in water. In the course of only a few hours, the mounting pressure of the swelling seeds has forced apart the interlocking joints. Undamaged, the skull falls apart into its different elements. The power that is locked in every seed, the power of plant growth that is now evident everywhere I walk on this late-March afternoon, is thus harnessed to assist in research.

MARCH · 27

THE PUGNACIOUS DOVE Mourning doves have been back for some time now, dropping into the yard with fluttering wings, helicopter-wise, to feed on scattered grain in the driveway. Watching them, we have become a little dubious about the dove as a symbol of peace.

One mourning dove in particular is a huffy-puffy bird that backs down for nothing. Several times, we have seen it inflate itself and walk with determined gait straight up to a bluejay. Each time, the jay has flown away. This morning, one of the gray squirrels hopped about the driveway, also feeding on the grain. The dove headed in its direction. When the squirrel showed no sign of giving ground, the pugnacious bird whacked it on the nose and sent it scurrying.

MARCH · 28

A STARLING QUACKS LIKE A DUCK During these spring days, the repertoire of the starlings is increasing. It is in the breeding season that the mimicking of other birds reaches its peak. One male has been sitting on a limb of the silver maple tree today giving the calls of nearly a dozen different birds. It mimics such varied species as the crow, the towhee, the crested flycatcher, the catbird, the meadow lark and the white-throated sparrow. For several days, a child in the neighborhood has been intermittently blowing a shrill police whistle. Now the starling imitates the sound, a little softer in quality but unmistakable. Some individual starlings appear specially gifted in mimicry. I believe it is the same bird that, on occasions, will imitate the "wicky-wicky-wicky" of a flicker and follow it with a sound that seems to mimic the hammering of its bill on wood. As I walked under the starling's tree, this morning, the bird was adding two new imitations to its repertoire. One was the call of the killdeer. The other was one I had never heard any other starling imitate—the quacking of a flying mallard duck. So realistic was it that I scanned the sky. Then the duck voice came again—directly from the limb where the starling sat.

MARCH · 29

SQUIRREL SHOWERS In the afternoon, a short and sudden downpour drenches the trees. Then the sky clears and rain is followed by sunshine. Looking from my study window, I can follow the progress of Chippy, running along the branches from cedar tree to cedar tree, by successive showers of drops shaken free in her passage.

MARCH · 30

BIRD-WHILES In the spring of 1838, Ralph Waldo Emerson made this entry in his journal: "A Bird-While. In a natural chronometer, a bird-while may be admitted as one of the metres, since the space most of the wild birds will allow you to make your observations on them when they alight near you in the woods, is a pretty equal and familiar measure."

In the century and more since those words were written, birds have been subjected to increasing pressures and harassments. Perhaps today they are less confiding and at ease. For it has been my experience that wild birds vary widely in the time they will permit observation and in the ease with which they take alarm. Different species and different individuals within the species have different bird-whiles.

Even in a backyard, such differences can be observed. English sparrows seem the quickest to take alarm; mourning doves the least ready to fly. It is interesting to note that recovery from fright is also a definite and rather constant characteristic. Scare up a yard full of birds when they are feeding and note the order of their return. This same order appears to be followed each time. Bird-whiles—going and coming—while not invariable are still an observable characteristic of all the birds around us.

MARCH · 31

A NEW LIGHT The light is everything. As I turned away from the lowland stream late this afternoon, and walked up through the Insect Garden and the old orchard, I looked back and caught the greens of the new grass clumps, the different shades of the young sweet flag leaves, all suddenly brilliant in the sun which had just emerged from behind a cloud. As in a vision, the greens stepped forward. Sometimes before a summer thunderstorm, we notice how the peculiar illumination brings out special details and alters a whole landscape. A change in light gives new character to a scene as surely as colored spotlights produce dramatic effects on the stage. As I stood there, the old saying took on added meaning: "To see it in a *new* light."

CHAPTER FOUR

April

APRIL • 1

NIGHT HERONS A wind from the south is raking the length of the swamp this morning. It is whipping the cattails, beating out the fluff, sending it streaming away in little puffs and clouds. As I watch, eight black-crowned night herons sweep past, turn, and parachute down into the tops of the tupelo trees across the swamp. They seem tired. With heads pulled in and shoulders hunched, they remain motionless most of the morning. This is migration's end. The travelers are home. Each year, around the first of April, the black-crowns come back. For three years in a row, when I first began keeping track of them, they reappeared on April 2. I began to wonder if they came back every year on exactly the same day. But later records have shown that, like most birds, even the swallows of San Capistrano, the black-crowns vary their arrival according to the spring—according to whether it is early or late.

APRIL • 2

APRIL WIND Yesterday's wind increased in the night. All morning it has been pounding in from the sea, a kind of wild, tumbling surf of the air. This afternoon, I walk across the moor of the wide salt meadows to the bay. I labor into gusts that buffet, sting, deafen me. They bring tears to my eyes. Wherever I go on this shelterless plain, the wind is pounding. It is whipping the dry cordgrass, lashing the phragmites, smiting the dark bay water, holding the gulls stationary above the mud of the beach at low tide. The gusts worry the seaweed hanging on the edge of the land where moor and water meet. They hurl themselves against the massed blue shells of the mussels. Seaweed and mussels, these two form natural buffers against the waves, protecting the edge of the land from erosion. Working against them, however, are the holes of the fiddler crabs. I see them all around. They riddle the banks, weaken them, lay them open to the work of the water. Wind and wave and fiddler crab—thus the mighty and the frail work together. Along the edge of land, big and little pieces move in an endless game of change and counterchange. Coming home, with the wind at my back, I am borne along like a lightly laden ship. The comparison is heightened by the tumbling waves of cordgrass rising and falling around me.

APRIL • 3

FREEDOM The freedom of the Red Man, the liberty of the child of nature, the unfettered life of the savage—these represent one of the great popular illusions of history. Everywhere and every time, where two or more people have lived together, mutual loss of liberty has resulted. For the good of all, concessions have been essential. Moreover, taboos and superstitions hemmed in the savage, restricted his freedom of action. He was, in his way, as fettered by conventions as is the man of today. Freedom from worries and surcease from strain are illusions that always inhabit the distance.

HOMECOMING SWALLOWS The wind of the past two days is gone, the sky is clear, and all across the surface of Milburn Pond tree swallows are shuttling back and forth, rising and descending, twisting and turning, feeding on flying insects. One passes within reach of my hand as I shade my eyes in the sun. Like the night herons, the swallows have come home. Perhaps they rode the great winds of the previous days, covering long distance with little effort.

APRIL • 4

WORLD'S MOST VALUABLE ANIMAL A long soaking rain before daybreak. And all along the way, when I walk for the morning papers, I find earthworms here and there stranded on the inhospitable cement of the sidewalk. Appearing naked and bewildered, they are in imminent danger of the early bird or of drying out in the sun-warmed air. This is the time of year when my morning walk is slowed by stops to put earthworms back on the ground where they belong. People probably wonder what treasure I am finding when they see me stoop so often!

And, in a way, I am dealing in treasure.

The pelt of a silver fox may sell for hundreds of dollars. The legs of a racehorse may be insured for a quarter of a million. Yet neither the silver fox nor the racehorse is the world's most valuable animal. This is the earthworm—a creature without fur and without legs; a creature that has neither jaws nor eyes nor ears; a humble burrower, nature's plowman.

As frost has left the topsoil, earthworms have worked upward. They have begun once more their invaluable activity of plowing, pulverizing, aerating, fertilizing, leveling and draining the soil. As many as 5,000 earthworms may plow through the earth of a single acre. During spring and summer and fall, they may bring as much as eighteen tons of new earth to the surface.

This labor is achieved with curious but effective equipment. Instead of legs, the earthworm employs hundreds of stiff bristles. Each body segment, except the first and last, is equipped with eight of these bristle-hooks. They are used to grip the soil on all sides and they explain the tugging of the robin when it seeks to drag an earthworm from the ground.

Underground, the earthworm pushes or literally eats its way through the soil. Its mouth, functioning like a suction pump, draws earth into its body. There it is pulverized and organic particles are digested. The rest is deposited as castings at the mouth of the burrow. In its surface feeding, during the hours of darkness, the creature usually anchors its tail in its burrow and then, elongating its body, moves in a circle like a tethered calf in its search for bits of decayed leaves.

At such times, it is warned of danger by curious senses, amazingly keen. Although it has no eyes, its skin is so sensitive to light that it warns the worm when dawn is breaking. Although it has no ears, it is so sensitive to earth vibrations that it is alarmed by the footfall of even an approaching shrew.

This is the earthworm, that humble and invaluable creature I see so frequently along my way this morning.

APRIL · 5

THE BLUEJAY WEIGHS THE PEANUT This morning, I threw half a dozen peanuts into the yard from the back door. A single bluejay dropped down from the maple tree. When several bluejays alight, each snatches a nut and flies away. But when one finds itself alone with several peanuts, it always does what this jay now did. It hopped to one nut, picked it up in its bill, then dropped it. It hopped to the next and repeated the performance. In this manner, it made the rounds. Then it returned to one of the nuts, snatched it up and flew away. Apparently the bluejay had tested all the nuts, weighed them in its bill, before making its choice of the heaviest one. Sometimes I have seen a jay go back to three or four of the nuts a second time before coming to a decision.

APRIL · 6

RETURNS OF SPRING Killdeer are flying over the open fields at dawn. The golden disk of an early dandelion shines beside the driveway. Greater yellowlegs and piping plovers have come back to the edge of the bay. And where the swamp path winds through Dragonfly Hollow at the foot of my Insect Garden, I come upon the probings of snipe. For a dozen paces the ground is riddled with holes. Here and there, the dry grass and refuse have been pushed aside by the working of a snipe's bill until it has produced a circle an inch or two in diameter. Everywhere I go there are evidences of life reappearing in its season. These are the returns of spring. These are fragments of the renewed life of the out-of-doors. It was gone and it has come again.

APRIL · 7

THE COMITY OF SPIDERS I have been reading a book on natural history, two centuries old, translated from the French. There is the charm of another time about such chapters as "The Splendor of Butterflies" and "The Comity of the Spiders."

DAWN LIGHT The dawn light at the window comes earlier but the robin is earlier still.

APRIL · 8

PREDATORS We talk of predators from a curious, restricted viewpoint. Is a robin a predator? To the earthworm, yes. Is the barn swallow a bird of prey? To the flying insects it feeds upon, certainly. The cuckoo that devours the tent caterpillar is no bird of prey but the hawk that devours the cuckoo is! As American jurisprudence is based on British law, so much of our thinking of "vermin" and "predators" is founded on the outlook of the British gamekeeper. He spent his life stamping out ruthlessly everything that interfered with the production of gamebirds. Nature has other goals in mind. Tent caterpillar and cuckoo and hawk all have their places in the scheme of things. The interplay of appetites has kept the world of nature on a level keel these many aeons. If man can take care of man, nature can take care of the rest.

APRIL • 9

SUPERSTITIONS ABOUT INSECTS I have found the pages of an old book, written during the Civil War, a repository for credulous ideas and superstitions regarding insects.

Do you want to know whether you will get a letter, be struck by lightning, break all your dishes or find a pot of gold? Would you like to learn if a suitor is faithful or if you will get a new sweetheart or if you will have a large family? All you have to do, the superstitious of many ages have believed, is consult the insects. They can tell you.

Step on a cricket, one superstition has it, and rain is sure to follow. Catch the first butterfly you see in spring, another warns, and you will be unlucky all summer. Kill a lightning bug, a third declares, and you will be struck by a thunderbolt the next time it storms.

Other folklore beliefs advise you to use care in dealing with many of the six-legged. If you disturb yellow butterflies clustered at a roadside puddle, you will lose a pot of gold. If you knock down a mud-dauber's nest, you will break all your dishes. If you see a swarm of honeybees alight on a dead branch in a tree, it means you will have a death in the family. To dream of ants is a sign you will have a large family or move to a big city. Dragonflies are supposed to spend part of their time sewing up the mouths of scolding women, saucy children and profane men. And bees are often engaged in punishing the wrongdoers. Peasant girls of Central Europe used to lead their lovers past beehives. If they had been unfaithful, the bees were supposed to rush out and sting them.

A number of insects have been considered omens of good luck. If a butterfly lands on your hand, it means you will get money; if it lands on your shoulder, it means you will get new clothes; if it lands on your head, it means you will have a new sweetheart. A bee circling around your head is supposed to indicate a letter is on its way, while a fly buzzing about you is an indication that a stranger wishes to meet you. Count the number of spots on the back of a hibernating ladybug and you will find as many dollars as there are spots. Hear a cricket in the house or let out a fly buzzing along the windowpane, and you will enjoy good fortune. If a bumblebee flies into the house in the morning, it means good luck; if it comes in the afternoon it means bad.

Such are some of the strange, persisting, unreasoning beliefs that have been associated with familiar insects in generations past. Credulity, in fact, has been the powerful ally of certain insects. The belief that killing a ladybug will bring bad luck has saved more of these little beetles than all the statistics showing their value to the farmer. The best protection in the world for a wild creature may be a superstition!

APRIL · 10

THE ONE-LEGGED BLACKBIRD One-Leg is back. He came in this morning, as jaunty as ever, and is now hopping about and holding his own among the other redwings feeding in the yard. This is the third consecutive year that One-Leg has appeared. He seems only slightly inconvenienced by the absence of his left leg. Tilting a little to the right, he has adjusted his body balance so he manages well on the ground. He has compensated for his loss in numerous ways. It is interesting to observe how he lands. Always at the last moment he sideslips to the right so he touches the ground with a certain amount of movement in the direction of his remaining leg. Often, in a wind, he comes in with a sideslip that reduces his ground speed almost to zero so that he alights with hardly more than a single hop along the ground. I never see him come to earth heading directly forward. Always he comes in with a slip that gives him momentum away from the side where the leg is missing.

We recognize him almost as much by his actions as by the absence of his left leg. Watching our backyard birds day after day, Nellie and I have come to recognize certain characteristics and mannerisms that identify them as individuals. Their actions mark them almost as surely as a numbered band would do. In this connection, I remember the celebrated case of "Gull Dick" as it was reported from time to time, years ago, in *The Auk*.

For twenty-four years, this herring gull spent his winters near the Benton's Reef Lightship, in Narragansett Bay, on the coast of Massachusetts. He would appear about October 12 and disappear about April 7. The men aboard the light-ship fed the gull regularly with bits of boiled pork and fish. At mealtime, in response to a call or a waving hand, Gull Dick would come flying. He would drive away other gulls, then swoop close beside the vessel to snatch bits of food. By his voice, his markings, his disposition and his actions, this bird was easily recognized from all other gulls of the bay during the more than two decades it returned to the light-ship.

APRIL • 11

THE PRESENT The present days are never like the days of old because we are never again exactly as we were in the days of old.

APRIL • 12

BIRD DETECTIVES We feel like scientific detectives this morning. We have found a clue that solves a mystery that has puzzled us for some time. On the wide sea meadows south of us, three-quarters of a mile or so away, we see starlings alight to feed where meadow larks alone fed before the first starling arrived from abroad in the late 1890's. Are any of the starlings we see there, starlings that come to our backyard? The answer is yes, positively. Our birds seem to disappear after they feed in the morning. Apparently we have definite proof they go to the meadows. They now are imitating shorebirds, greater yellowlegs, plover, and others, that they hear along the edge of the bay in the salt meadows. This deduction is no more difficult than the bit of bird-detective work that identified a small pure white bird near Washington, D.C., a few years ago. It was put down as an albino white-throated sparrow. But, as the white patch on the throat of this sparrow is its chief field mark, how would anyone know it had a white patch if the bird were all white? The answer was simple. The bird was a male and it sang the song of the whitethroat.

APRIL • 13

MARSH MARIGOLD To Mill Neck to see the marigold. All along the little streams, in the lushness of spring, the sappy lowland plants, false hellebore, skunk cabbage, marsh marigold, are massed together. In mounds, the marigold lift their shining, waxen-petaled flowers. Overhead, where the baby leaves have pushed out from buds, the sky is speckled and dappled, with sunshine pouring down on the black, wet soil. The quiet of the morning, broken by the low water music of the rill and the occasional thin lisping of a golden-crowned kinglet, this quiet seems so intense that it is only a short step to the fancy that we can hear the plants pushing upward, unfolding, multiplying cells, all around us. We feel like the character in Grimm's fairy tale who put his head to the ground and heard the grass grow.

APRIL • 14

RAIN One of the last whitethroats of the season is singing in the misty rain. And, under the trees, where it perches, their wide leaves sprinkled with shining drops of water, a clump of purple violets bloom—the first I have seen this spring.

APRIL • 15

SPRING BUMBLEBEES At 6:30 this morning, I watch a velvet-coated bumblebee begin her day of hunting for a nest site. In five minutes, as I follow her along the lower slope of the Insect Garden, I see her investigate every possible opening near a pile of moldering fence rails. Zigzagging, hovering, circling, alighting, she covers the slope as thoroughly as a man could do it. I see her alight on a fallen paper and peer under the edges. She explores under a maple root, beneath dead leaves, in four grass clumps, in a knot-hole in a log, by a weathered glass jar in the weeds, under a fallen branch, in the region of my shoe, at the foot of a cedar tree, along a bit of board lying in the grass. In five minutes, she investigates thirteen possible sites. Multiply this by the hours and days she will continue her searching and we have some idea of the hundreds and even thousands of sites she considers before she comes to the important decision of where she will establish her nest.

The overwintering queen is the founder of an insect city. The fate of the colony depends to a great extent upon the wisdom of her choice. I have watched other furry, black-and-gold bumblebees disappear into mole tunnels, investigate rusting tin cans, seek out the underground, abandoned nests of meadow mice. Often, the final choice is a mouse nest. There the young queen builds a thimble-shaped honeypot from wax exuded by her own body. This she fills with a mixture of nectar and pollen. Then she forms waxen cells, places at the bottom of each a layer of pollen and on it deposits an egg. Covering over the cells with lids of wax, she guards them, sitting on them almost like a brooding hen, leaving them only for quick trips afield for food.

Out of each egg will come a pollen-eating larva. By the time the mattress of pollen is gone, the larva is full grown and ready to spin the papery cocoon within which it transforms into the adult bumblebee. When they first appear, the young bumblebees are drab gray. Like goldfish, they get their brilliant yellow coloring later on.

The first-born of a colony are always workers. They build cells, gather food and care for the nest, leaving the queen free to lay hundreds of eggs that increase the colony's population day by day as the season advances. At its peak, in late summer, one of these insect cities may have a total of as many as 500, or even 1,000, bumblebees. Toward the end of its life, the males and the young queens appear and mate. Then, in the cold of autumn and winter, all the bumblebees will die except the fertilized queens. They will remain hidden in hollow logs, under rubbish, in secret places, until the coming of spring awakens them. Then again, as on this sunny April morning, each awakened queen will hum on her winding, zigzag path low above the ground, searching out the place where she will establish and populate the bumblebee city of still another year.

APRIL · 16

COTTONTAILS OF THE AIR Under the overarching maples and wild cherries at the southern end of the swamp path, I notice anew this morning how the white rump-patches of the flickers shine out as they go galloping ahead of me down the trail. The flicker is one of the cottontails of the air. What are the others? The marsh hawk, the greater yellowlegs, the bald eagle, the white-tailed kite, to mention four.

APRIL · 17

DAWN AT MILBURN POND Soon after daybreak with Nellie to Milburn Pond. It might well be called Mallard Pond. There are always mallards on it. This morning we see the first of the ducklings, twelve baby mallards, apparently just out of the nest, darting about their mother like active little waterbugs. A pair of swans is busy building a nest, dipping long necks to the bottom to pull up waterweeds which they add to the pile. Along the opposite shore, as the morning advances, two redwings battle in the air, rising up and up, then dropping, still clawing and pecking until they both hit the water with a splash. Only then do they part and fly to different trees. Most of the redwing fighting is over now, territories are well established apparently, and before long, nest-building will be in full swing. On our way back, we stop to watch a muskrat feeding at the surface in the sunshine. Above us, in a wild cherry tree, a yellow palm warbler bobs its tail and turns its head to watch us as we walk on. It reminds us that the largest waves of migrations, with all their variety and excitement and interest, are close at hand.

APRIL · 18

THE APPETITE OF A SQUIRREL This morning, the redoubtable Chippy sat on the kitchen window ledge and consumed nuts almost as fast as I could hand them to her. Without going away and without burying any of the nuts, she consumed, in hardly more than half an hour, the following: twenty unshelled peanuts, virtually all of them containing two nuts, two pecans, one Brazil nut and three almonds, a total of at least forty nuts at a sitting! It may be that her food requirements are abnormal at present as she seems to be nursing a brood of baby squirrels in the hollow silver maple tree.

APRIL · 19

A STARLING SWINGS A RIVAL BY THE LEG A pair of starlings have taken over the nesting box on the grapevine trellis. This morning, there was a great commotion at the box. A stranger starling had tried to enter while one of the owners was at the nest. By the time we trained our glasses on the spot, the owner had its bill clamped on one leg of the interloper and was jerking and swinging its rival this way and that. The moment it could break free, the intruder fled pell-mell in squawking terror over a garage, beyond maple trees and out of sight.

APRIL SUNSHINE At Milburn Pond, in the April sunshine, a painted turtle has clambered up the side of the swan's nest and is resting there, its neck outstretched, its body soaking up the warmth, its sluggish winter blood reviving in the spring. It is between one swan sitting on the nest and the other pulling up waterweeds and adding them to the pile. Neither pays any attention to the turtle. The blackbirds are more quiet now. They seem to be settling down. They are less noisy and more serious—as befits persons of property.

APRIL · 20

THE SONG OF THE WOODCOCK Now, in the April twilight, the woodcock sings. We go to the cedar woods this evening to listen to this rare and charming music of the Spring.

Shy and silent throughout the rest of the year, the male now mounts from the accumulating shadows in fields beside the woods, circles upward into the luminous sky, then comes tumbling down, filling the dusk with the sweet, twittering song of his courtship. Other wild singers have greater vocal range, more virtuosity. But the strains of very few birds are as moving as this simple, liquid song of the woodcock heard amid the hush of the darkening woods and fields.

First we hear, issuing from the shadows of the field, a muffled, gurgling sound, then a nasal, buzzing "peeent!" This is repeated at intervals of from two to a dozen seconds, with occasionally a longer pause between the calls. Then comes a sibilant, winnowing sound. The woodcock is in the air and we follow its dark little form speeding across the luminous sky of the western horizon. It swings in great circles, up and up, until it is lost to sight.

Then down into the shadows comes a swallow-like twittering. It increases in tempo and for a moment we catch a glimpse of this strange little singer as it plunges out of the sky, gyrating like a windblown leaf, descending in wild, ecstatic flight and singing as it falls.

Just before it reaches the ground, it straightens out and, with a rustle of its wings, lands close to the place where it took off. Near by, hidden among vines and bushes, is the female for whose eyes and ears the performance is intended. So absorbed is the male in this exhibition that I run out on the field as soon as he ascends and throw myself down on the ground near the spot where he arose and watch him tumbling down toward me to alight no more than half a dozen yards away.

In the deepening dusk, I turn the beam of a pocket flashlight upon him as he struts about uttering his nasal "peeent!" It disturbs him not in the least. But if I make even the slightest movement while he is on the ground, the shy bird will take fright and the twilight show will end.

The most memorable of these song flights I have observed near these cedar woods occurred one misty night. I lay on my back amid the dewberry vines of the open field. Overhead, the song of the woodcock descended through the mist, increasing in volume, until the bird, still singing, burst from the white vapor and glided down, with a soft winnowing of wings, close beside me.

This annual show, between the setting of the sun and the coming of the dark on these spring evenings, forms one of the most profoundly moving adventures of the year.

APRIL · 21

REFLECTIONS IN A STREAM When I start for the swamp walk in the misty dawn, this morning, the radio is collecting the troubles of the world and packing them into a five-minute broadcast. Woe and foreboding fill the air waves. As I switch off the set, I recall the famous line of the comedian, Eddie Cantor: "We are facing a crisis and the question is, what is the answer?" But out-of-doors the news is good. All of nature is a going concern. The business of Spring is prospering. I stand for a long time beside the swamp stream in a fairyland setting of low-lying mist glowing and tinted with the pink of the sunrise. Here is beauty and here is calm. A slight breeze stirs the phragmites and the reflections of their high, plumed heads move from side to side and at the same time continually lengthens and contracts—the lateral movement produced by the swaying of the grasses and the elongation and contraction by the slow ripples running across the dark mirror of the stream. Such sights as these set us to rights again. For the mind disturbed, the still beauty of the dawn is nature's finest balm.

SALAMANDER EGGS This afternoon, I stop in to see Edmund Morgan, Curator of the Tackapausha Museum, at Seaford. He is watching the eggs of spotted salamanders hatching in an aquarium. Inside each little transparent globe of an egg, we see the occupant, a minute, streamlined creature curved like a quarter moon against the invisible shell. We can distinguish its speckled body, its dotted tail, its miniature bunched gills. With the light coming strongly from one side, I photograph the eggs and the little prisoners within the transparent walls. Each ball of an egg seems a perfect sphere of crystal glass. As we watch, one after the other the midget salamander progeny hatch, and, with gills outspread, swim away—launched into life.

APRIL • 22

ENCHANTED ROBINS Near here, a robin has been fluttering against a window and pecking at the glass, hour after hour, all day long. At intervals, it alights in a cherry tree to rest. But the limb is near an upper-story window and in a few minutes it is back again, buffeting itself against the glass.

This is the time of robins bewitched. People call me on the telephone and write letters about robins that peck endlessly at shiny hubcaps on automobiles, that spend their days at window panes. One woman in Camden, N.J. writes that a robin there has been fighting with a mud puddle after a rainstorm. Have the birds gone crazy? Are they trying to break through the glass? What ails them?

All over the country, where robins are nesting, the same thing is taking place. Males are defending their nesting territories, flying to attack any interloper. Like many birds, robins have definite nesting and feeding areas. I recall an instance in which an ornithologist in New Jersey once found that one pair of robins held the feeding rights to the ground beneath a tree while another pair of robins held the nesting rights in the tree itself. Neither pair encroached on the domain of the other pair. The territory-sense among birds serves them well. It spreads out the species and prevents overcrowding in one area.

Catching sight of its own reflection in a window or on a shiny metal surface, the male robin dashes to drive the intruder away. He may battle this phantom rival for days on end. A friend of mine, who was banding birds, once found a robin fighting with its reflection in a basement window. He caught the bird, banded it, and released it a mile or so away. When he returned to the building, at the very same window a robin was pecking and buffeting the glass. While the first bird was being transported away, a second male had caught sight of his reflection and had flown to the attack.

Only windows with darkened rooms behind them, turning the glass into a mirror, attract the attention of the birds. Consequently, a bewitched robin can be freed from its spell merely by turning on a light in the room or hanging a white cloth in the window. This will break the enchantment. The bird's reflection will disappear. The robin will imagine it has vanquished its rival. In high good spirits, it will return to the normal life of a redbreast in the spring.

APRIL · 23

SURROUNDINGS I talked to a scientist today who is experimenting with primitive forms of sea life. Even the lowliest creature, he has found, reacts differently in a tank than in its natural surroundings. Change the surroundings and you change the animal. We talk of animals being part of their surroundings; we might also note that the surroundings are part of the animals as well! In all tests where living creatures are studied in the laboratory, the effect of artificial conditions is a factor that cannot be ignored.

APRIL · 24

BREAD UNDER A LEAF The ways of the bluejay are a source of never-ending interest. Every year, people write me asking how they can get rid of the bluejays. Getting rid of the bluejays at a backyard feeding station is getting rid of much of the interest there. Just now, one of these birds has picked up a crust of bread and flown across the yard to the wall of a neighboring garage. There it has carefully laid down the bread and picked up a weathered maple leaf that fell to the ground last fall. This it places on top of the bread, hiding it from the sight of squirrels and other birds. Like the gray squirrel that covered the peanut butter with the shells of sunflower seeds, the jay is laying by food for another day.

APRIL · 25

GREEN FLOWERS In clumps and clusters, green flowers are blooming all over the Norway maples. First the flowers, then the leaves. The pear trees are white clouds of blossoms but the bracts of the dogwood are still green. Each morning, I awake to the rapping of flickers on hollow, resounding wood.

APRIL · 26

A FLICKER ATTACKS A STARLING It is now more than half a century since the first European starlings were released in Central Park in New York City. They have multiplied and spread until they have reached the Pacific Coast and Churchill, on Hudson Bay, and Mexico City. They have taken over nesting territory from bluebirds and flickers. On our part of Long Island, bluebirds no longer nest and I have seen starlings perched on limbs of a tree waiting until a flicker finished making its nesting hole which they then appropriated. Twice I have found holes cut in trees at the bottom of a flicker hole, thus rendering the nests useless to the starlings. This seemed to be a form of flicker revenge. But I have never seen one of the big woodpeckers make a direct attack upon a starling until today. This afternoon a female flicker landed in our yard to feed upon ants. As I watched it from a window, a starling came walking up with its characteristic swinging swagger and pushed close to see what the flicker was finding in the grass. It walked completely around the feeding woodpecker at a distance of hardly a foot. As it came closer from the rear, the flicker suddenly leaped like a cat. A hop or two and it was hammering the starling in a flurry of wings that suggested a cockfight. In a moment, the melee was over. The starling flew away squawking and the flicker returned to its ant-hunting. Perhaps, in the end, flickers will learn how to cope with starlings.

APRIL · 27

THE BROWN THRASHER'S SONG A brown thrasher came in during the night. I hear it singing—going on and on with starts and stops, with exclamations and asides—when I awaken this morning. It is singing again at sunset. Listening, I remember that as a small boy I heard this song translated into words. I believe it was the first time I ever encountered words used to represent a bird song. The brown thrasher, I was told, says this as it sings:

"How much will you give me for my tail—my tail? A shilling? A shilling? 'Taint enough! 'Taint enough! A quarter? A quarter? Cut it off! Cut it off!"

To Henry Thoreau, when he was planting his bean field near his hut at Walden Pond, the brown thrasher seemed to be calling different words in at least part of its song. It seemed advising: "Drop it, drop it,—cover it up, cover it up,—pull it up, pull it up, pull it up."

APRIL · 28

LOVE OF WILD ANIMALS There is always a greater love. Those who wish to pet and baby wild animals, "love" them. But those who respect their natures and wish to let them live normal lives, love them more.

APRIL · 29

FALLEN PETALS Rain in the night and the fallen white petals of the pear trees lie scattered across my path like confetti when I walk for the morning papers. Mingled with the gray rain has been the green rain of descending maple flowers. They dot the sidewalks and form yellow-green windrows at the edges of the puddles. Around the larger pools, they seem like seaweed thrown up by the waves on the shore of a bay.

Leaf-fall in the autumn and flower-fall in the spring!

APRIL · 30

DAYBREAK SONGS Now all the days have music in them. The time of the singing of birds—it commences with the earliest dawn. I awake to bird-music. This morning, I noted the sequence of songs and the hour, Eastern Standard Time, when they began.

3:30 A.M. The robin—the earliest singer of all—begins in the faint, gray light of the first beginning of dawn.

4:25 A.M. I hear a bluejay call. It gives its musical bell note, not its harsh scream that sounds like "Thief! Thief!" That comes later in the day.

4:50 A.M. The brown thrasher, on a fence post, begins its long repertoire.

4:55 A.M. The first trilling song-sparrow song of the day.

5:05 A.M. A towhee calls. Yesterday we saw the first chewink of the year hopping about, kicking up fallen leaves, in the rose garden.

These are the first singers of the dawn on this final day of the month of April.

TOAD MUSIC Walking beside the swamp at 3:30 P.M., I heard music of a different kind. There came to my ears, for the first time this year, the metallic, far-carrying bray of a Fowler's toad. The spring peeper, *Hyla crucifer*, has been filling the marsh-nights with sound since early April. A few American toads, during the past week or so, have been lifting their voices in that sweet, sustained trill that is held for half a minute at a time. Now begins the shorter song of *Bufo fowleri*. Each year, these batrachian love songs of spring begin in the same order: First the peeper, then the American toad, and finally *fowleri*.

CHAPTER FIVE

May

MAY · 1

THE DAWN OF MAY Another dawn—serene and honey sweet. At such times, it seems to me that dawn is nine-tenths of the day. Staying up late at night has a sameness about it; but every dawn is different. And this is the dawn of May—May, the month that is never long enough. This is May the first as the first of May should be.

I saunter along the swamp edge and beside Milburn Pond, where violets bloom. A killdeer circles and calls in the open field behind me. Overhead, I hear the winnowing wings of a mourning dove passing by, and, out among the cattails, a female redwing plucks fluffy down from one of the weathered heads and flies away. Brilliant green runs across all the tussock tops at the edge of the flooded lowland.

On such a day as this, it is enough to spend the hours soaking in the sunshine, breathing slowly, sensing to the full all the perfumes of spring. It is enough to delight in the varied shades of green, in the forms of trees and the colors of flowers. On such a day, all our moments out-of-doors are lived in quiet pleasure. I lean against one of the ancient apple trees on my Insect Garden hillside and words I once read on the title page of *The Fisherman's Bedside Book* come back to me: "The wonder of the world, the beauty and power, the shapes of things, their colors, lights and shades; these I saw. Look ye also while life lasts."

MAY • 2

WILLOW SPROUTS Beside the road, at the edge of the salt meadows, someone has dumped the limbs and trunk of a willow tree. The logs have been lying there a good part of the winter. Today, as I walked past, I noticed that sprouts have pushed out all along the length of one of the logs. I stooped to count them. The seven-foot log was about eight inches in diameter at one end and about four at the other. One hundred and fifty-six sprouts had pushed out from the wet wood of this discarded log. The sight recalled the experience of a gentleman of this vicinity who placed rustic chairs about his yard. He left them out all winter. In the spring, he discovered that every chair had taken root. The rustic wood was willow.

HAILSTONES Late in the afternoon, at the conclusion of a day of intermittent showers, the clouds darkened, a sprinkle of rain descended, and then, for the space of several minutes, hail streamed from the windless sky. I was crossing the yard at the time and the little balls of ice, white and shining, bounded off my arms, my head, my shoulders. They lodged in the grass. They rolled on the wet ground and lay there, gleaming in the murky light. During the period these Maytime hailstones were falling, the mercury in the thermometer stood at 48 degrees F.

MAY • 3

THE OWL TREES Two black-and-white warblers, as brilliant as contrasting black and white can be, dart among the cedars outside my study window as I sit down to work this morning. A little after eight, an uproar of bluejays commences in one of the high, dense spruces in front of the house. It continues, increasing in volume, as new jays arrive. Deciding they must be mobbing an owl, Nellie and I go out and sweep the upper branches with our glasses. It is five minutes before we make out part of a gray-plumaged bird, then the heart-face of a barn owl. This is the first barn owl we have ever had in the yard. But in the other spruce, a dozen feet away, bluejays once called attention to a rarer visitor by far. It was about eleven o'clock in the morning and the clamor of the jays went on and on, minute after minute. We were busy, and only after ten minutes had gone by did we investigate the cause of the uproar. A great white bulk lifted out of the tree top and sailed away with steady beats of widespread wings. It was one of those rare and cyclical visitors from the tundras of the far north, a snowy owl.

Over the years, we have noted the birds seen in our yard. Once a woodcock crouched beneath the low limb of an arbor vitae tree near the front porch. Another time, a northern water thrush bobbed about on the ground beneath the "snowy owl" spruce. Hairy woodpeckers have alighted in the maple trees and, at various times, five kinds of thrushes have fed in the yard. Rarest of our avian visitors, with the exception of the snowy owl, was a dovekie, a little swallow-like pelagic bird of the Arctic seas. A few winters ago, a gale over the Atlantic drove exhausted dovekies inland. One came down in a tumbling, web-footed landing on the grass. In the end, we restored it to salt water near the Jones Beach Causeway. As Nellie said, the way to see rarer birds in the field is to bring them with you!

MAY · 4

THE SHYNESS OF NATURE Nature is shy and noncommittal in a crowd. To learn her secrets, visit her alone or with a single friend, at most. Everything evades you, everything hides, even your thoughts escape you, when you walk in a crowd.

CATBIRD NEST In the underbrush by Milburn Pond, this morning, we heard the mewing of a catbird—the first we have seen. A little later, we come upon it hopping about a last year's catbird nest. The thought that comes first to mind is that this is a fledgling returned, a bird of last summers' brood, visiting its old home, the nest where it was born. In truth, it is more likely a bird that has nested here before. It is the nestlings that scatter, that move into new breeding territory. It is the old birds, the parents, that tend to return to the spot where they have nested the previous year.

MAY • 5

A YODELING BLACKBIRD One-Leg, well and strong, sideslips in each morning to feed on the driveway. As in the previous two years, he is the most individual bird we see. But, this spring, another red-wing stands out by virtue of its call. Instead of the familiar "Okaleee!" it gives a kind of double yodel followed by a long-drawn trill. We recognize it the instant we hear it. It is banded by a call. This vocal peculiarity sets it apart from its fellows as clearly as does the loss of a leg for One-Leg or the white feathers in the plumage of a white-winged blackbird, a partial albino, that appeared one summer among the cattails of Milburn Swamp.

MAY • 6

THE SEASON ADVANCES I follow the old familiar paths amid endless change on these successive days of spring. In multiform ways, the season advances. Keys have formed on the maples and yellow-green flowers, like bursts of fireworks, spray out on the sassafras twig ends. The fiddleheads of the cinnamon fern are unrolling. Bellwort and cinquefoil and wild strawberries are in bloom. I walk beneath a post oak tree so laden with catkins it seems dripping with Spanish moss. It seems only yesterday that I saw a robin with mud on her breast, the consequence of her nest building. Now I find the brilliant blue of a hatched robin's egg beside the path. Aronia and Leucothia are in bud. I hear the first oriole song of the year this morning. Where dogwood bracts have been losing their green, day by day, they are now white, now nearing that whipped-cream brilliance of their mid-May glory. Pear blossoms are gone and cherry trees are already beginning to look tarnished and threadbare. But on the slope of my Insect Garden, above the swamp, the pink-white cumulus clouds of blossoms are beginning to billow up on the apple trees. A veery alighted in the yard today and warblers are increasing in the oak trees. At the edge of Milburn Pond, in the shallows, the first patch of cleared pebbles, the first nest of a sunfish, records also the advance of the season.

M A Y • 7

THE TURTLE HUNTER I met a man this morning splashing in hip boots among the cattails at the edge of Milburn Pond. He was prodding the mud with a long pole with a steel hook at the end. Every year he comes here, he told me, hunting for snapping turtles. His annual catch, made during early morning hours before his work day begins, runs from 400 to 600 pounds. He sells the turtles for soup to a buyer in New York. His hunting ground is the pond and the swamp stream that winds away among the cattails below my Insect Garden hillside. No doubt his hunting benefits the mallards of the pond whose ducklings often fall prey to the larger snappers. I see one female mallard swimming about this morning followed by but a single duckling of all the brood she hatched so short a time ago. My talk with the turtle hunter brings to mind a belief firmly held about snapping turtles in the dune country of northern Indiana when I was a boy. I was solemnly assured that different parts of such a turtle represented the meat of various animals so that if I ate a snapping turtle, I would be dining on venison, bear meat, beef and mutton!

M A Y • 8

A DUCK ATTACKS A SWAN As we came through the fringing of woods to the shore of Milburn Pond at sunrise this morning, we heard one of the mute swans take off and land, unseen, behind the trees. First, there was the heavy "flap-flap-flap" of his wings smiting the water; then a whistling winnow as they moved in air; then a buzzing sound as his feet trailed through waterweed; then the rapid "pat-pat-pat" of his flat feet striking the water and, finally, a rushing splash as he landed.

The cob, these days, is patrolling the center of the pond, challenging everything that comes near the great pile of waterweeds where the female is nesting. Once we saw three stranger swans come overhead, circle the pond with whistling wing beats, and fly away. All the time they were above the water, the male cruised angrily about with ruffled feathers. He flies to the attack whenever any of a flock of half a dozen tame white ducks ventures out from the western shore. This morning, we saw him send flying a coot, a blue-winged teal and several male mallards.

About seven o'clock, a female mallard headed across the pond closely followed by a little flotilla of twelve ducklings. Her course took her close to the nest of the mute swans. As she approached with her little paddling brood, the male swan came charging down the lake. His bill was open and his neck outstretched. But, instead of fleeing, the mother duck darted to the attack. She flew in, with wide-open bill, first from one side, then from the other. Her fluttering, quacking onslaught seemed to disconcert the swan. He churned about, rushing and turning without being able to come to grips with his antagonist. Half a dozen times, the mallard darted in and swerved away. All the while, the flotilla of ducklings swam steadily away toward the shore and safety. With one final rush and swerve, the plucky mother mallard turned away from her ruffled antagonist and swam swiftly to rejoin her brood.

MAY • 9

"IN A MERY MORNYNGE" It was of such a day as this that the words, so long ago, were written: "He came to a grene wode in a mery mornynge." From the lowly plantain, now in bloom with its tiny white flowers encircling its head like the whirling bodies of a miniature solar system, to the gulls that fly with rowing wing beats overhead and the

painted turtles, sunning themselves on stranded refuse in the sw
stream, there is life and health and growth under the warmth of
sunshine. The perfume of apple blossoms is in the morning air and, out in the swamp, that tiny, bursting bundle of life, a long-billed marsh wren, clatters into the air above the cattails. All the leaves of tree and plant are new and perfect. Everything seems as immaculate and fresh as though I walked the earth on a first day of creation. Wherever I go, I experience the emotions that once prompted an old man to confide: "On a morning like this, I take off my hat to the beauty of the world."

MAY • 10

SOMETHING WHITE IN A NEST Walking home with the morning papers today, I heard a fluttering rustle in the air behind me. As I turned, an English sparrow passed me and trailed before my eyes a long streamer of tissue paper fully two and a half feet long and an inch and a half wide. It looked like a dark locomotive at the head of a white train as, with laboring wings, it sought to drag its lengthy burden upward toward its nest. But the resistance of the trailing tissue paper was too great. The bird could hardly maintain itself in horizontal flight. At last it gave up, let go, and the dropped strip of paper floated down almost at my feet.

It is an interesting thing to note the number of birds that show an inclination to add something white to their nests. A correspondent of mine, living in the country near Oxford, Ohio, writes of an experience with orioles. She was in the habit of placing strings of various colors on her clothesline when the orioles were building their nests. She noticed they always left the red strings and chose the white ones first of all. Once she tied a red piece and a white piece together and hung the white end over the line. The oriole took it and flew to the site where she was building her nest. But, in the autumn, when the brood was gone and the nest was obtained from the tree, it was discovered that the white string had been interwoven into the nest while the red end had been left hanging outside.

John Burroughs, in one of his books, writes: "I am at a loss to know why certain birds have such a penchant for something white woven into or placed on the outside of their nests. A robin will reject bits of colored paper, but will often use strips of white paper or white rags. One in the vines of a near-by shed has made very free use of the castoff hair of our old gray mare, nearly white. On a friend's house, in a Michigan city, I saw more than a yard of candlewick dangling from an unfinished nest. Nearly all the vireos have a habit of sticking bits of white material on the outside of their nests, usually the weavings or cocoons of spiders."

Once, in an elm tree on a village street, Burroughs watched a yellow-throated vireo finishing her nest. Seeking something white to place on the outside, she discovered petals of roses which had fallen to the ground. Time and again, she attempted to attach petals to her nest. They were smooth and, unlike the woolly or sticky spider silk, refused to adhere. Yet she kept on trying, over and over again.

Only the other day, a couple of miles from here, I was shown a fishline, thirty feet long, entangled in the branches of a maple tree. The line had been laid out on the grass to dry and a robin had tried to carry it—all thirty feet of it—upward to where the bird was building a nest in the tree top. This was the second year in a row that the owner of the fishline has had to climb a tree to retrieve it from the robins intent on adding white string to their nest.

Every year, far and near, the great crested flycatcher searches until it finds the light-colored discarded skin of a snake which it uses in its nest building. During one recent summer, a pair of these birds, unable to find a real snakeskin, adopted a long strip of cellophane as a substitute. In another instance, a pair of catbirds made off with two lace collars that had been laid out on a lawn to dry. They added the white lace to their nest in a near-by thicket. And when a pair of brown thrashers set up housekeeping in the backyard of a friend of mine, they brought not only twigs and rootlets, but a fragment of *The New York Times*. This was carefully woven into the side of the structure. There it remained all summer long, adding to the nest its touch of white.

MAY · 11

AN ORIOLE'S ADVENTURE Two Baltimore orioles, males in full breeding plumage, hopped from limb to limb in an oak tree in the Milburn Woods soon after sunrise this morning. Nellie and I watched them face each other in some kind of rivalry or display. Suddenly both birds sprang into the air and descended, fighting in a flutter of wings. Two feet from the ground, they parted. One flew away. But the other hung in the air, remaining in one place and fluttering as though fighting a phantom rival. I came closer and saw that it was caught by the neck where two cat briars crossed. I rushed to save it but just before I reached it, it tore itself free, apparently unharmed, and darted away.

MAY · 12

SONG OF THE SANDPIPER Warblers have come in in a wave during the night. In the Milburn Woods, they dart among the infant leaves of the oaks—parula warblers and myrtle warblers, chestnut-sided warblers, magnolias and yellows and black-throated greens and, a great rarity among these trees, a bay-breasted warbler with the sun full upon its deep red cap, its bay sides and creamy buff patches down its face and throat. A perfect male in perfect light!

Beyond, at the edge of the pond, we come upon the courageous mallard and her dozen ducklings resting on the top of a muskrat house. They skitter away down the side of the mound and out from shore. The spots of cleared gravel, the nests of the sunfish, are increasing in number. Robins alight for mud beside the pond and a little flock of a dozen or more least sandpipers feed ahead of us, taking off, landing and taking off again as we follow the shore. As they feed, they often give a delicate musical trill, somewhat reminiscent of the spring call of the sora rail, only more diminutive and finer in quality. At intervals, two of the birds will leap into the air and fly toward each other. The sandpipers are returning to their breeding grounds and the days of courtship are near at hand. One sandpiper, in particular, sings frequently. We listen in delight to its frail, sweet, lacelike little song.

MAY · 13

BOTANICAL LIZARD TAILS At the Insect Garden, toward sunset, I spade up some of the ground between the ancient apple trees for another year's planting of sunflowers and nicotianas and marigolds and cosmos and petunias and morning glories and other annuals that will help attract varied insects to the hillside. The perennials, the buddleia and buttonball bushes, the roses on their trellis and the honeysuckle vines among the wild cherries, have all come to life with the Spring. I notice how maple branches that broke away in winter storms and fell at the edge of the creek have produced their leaves in season. The branch of an apple tree, that came down in a December gale, is bearing blossoms. And twigs that I trimmed from one of the forsythia bushes, and discarded at the swamp edge, have been yellow with flowers. Even though severed from the living bush and tree, these fragments have continued, for a time, to live on the food stored up in the winter buds. They seem to be the botanical counterpart of the discarded lizard tail that temporarily retains the activity of life even after being separated from the source of life.

THE MATING OF THE TREE SWALLOWS High above the road, as I came home, I saw the spectacular mating of a pair of tree swallows. The birds flew wildly, almost like swifts. Then, fifty or sixty feet above the ground, the swallows met face to face, fluttered for a moment, and then, with wings wide spread to break their descent, they fell straight downward through the air. For forty feet or more they dropped before they disunited to rise up and up into the sunset sky. When they parted at the end of their aerial mating, the birds were hardly more than ten feet above the road.

MAY • 14

JOHN KIERAN'S WARBLER WAVE I cross the Whitestone Bridge at dawn for my annual morning with John Kieran in the midst of a warbler wave. We wander along paths through the wilder portions of Van Cortlandt Park, paths he has followed almost daily for more than thirty years. As always, when we go afield, it was a memorable and glorious day.

In this region, the northward movement of spring warblers is mainly along the Hudson River Valley. The band of their flyway does not extend far out on Long Island. Hence, I see far fewer warblers, far fewer species, twenty-five miles away than are seen close to the river. When we totaled up our list today, we found we had seen nineteen different kinds of warblers: the yellow, the prairie, the black-and-white, the myrtle, the parula, the magnolia, the chestnut-sided, the bay-breasted, the Canada, the redstart, the Wilson's, the blue-winged, the yellowthroat, the Blackburnian, the ovenbird, the northern water thrush, the black-throated blue, the blackpoll and the black-throated green. I think we both were a little sorry to see the blackpoll. For the blackpolls form the rear guard in the northward parade of the spring warblers. When they appear, the end of the warbler migration is close at hand.

MAY • 15

THE INSECT GARDEN Once more, today, as I have done for nearly fifteen years, I drop seeds, cover them up, set out plants, bring pails of water from the swamp stream, and thus start my Insect Garden on another cycle of its existence. From this activity will come leaves and flowers that will be hosts to the summer insects that I will watch with interest.

As I work in the fading light, honey-colored gnats drift past and I hear the pumping of a bittern across the swamp. The rich smell of the fresh earth is in the evening air. A mourning dove lands on one of the branches of an apple tree less than a hundred feet away. When it begins to call, even though I watch it so close at hand, its mellow, mournful cry seems coming from an immeasurable distance.

The planting done, I wander over the hillside in the deepening dusk. I note changes that have come with the years. Phragmites have increased in the swamp. Viburnum and wild cherry have spread on the land. The clump of golden asters, small when I first knew it, is now a dense island of plants several feet across. The wild bee tree has blown down and limbs have been lost by the decaying apple trees. The spot at the swamp edge where I photographed the tangle for the endpapers of *Grassroot Jungles* is now under water and overrun by the massed sword leaves of the sweet flags. Nothing stands still. No two years are alike. Everything that lives is constantly changing.

MAY · 16

SONG OF THE WOOD THRUSH Every evening, now, the wood thrush sings. Rich and rounded, its bell tones fill those two times of half light that form the boundaries of our day. Its song, like all earthly things, contains intimations of perfection limited by imperfection. The end of the song is something of an anticlimax. The glorious notes conclude in a kind of sniffling buzz. If the end came first, if the song soared to its perfection in a triumphant climax, how much more dramatic it would be! But even so, few wild singers in the world surpass the wonderful richness of the wood thrush's finest tones as they carry through the dusky woods. We soon forget the less-than-perfect ending for the nearly perfect song.

MAY · 17

A HEAVY-LIDDED DAWN At six-thirty this morning, the pond is glassy under a dull and rainy sky. Tree and barn swallows sweep back and forth almost touching the water. The mute swans are sleeping, one on the nest, the other near by. The dawn seems half-awake, heavy-lidded, unable to shake off the torpor of the night. For a long time, Nellie and Martha Meinke and I watch a whole family of six little gray squirrels feeding among the upper branches of a hollow oak tree. This may be, perhaps, one of their first trips outside the hollow trunk that is their home. Yet they climb sure-footedly among the gray new leaves and tassels on the twig tips, apparently feeding on parts of the catkins. They ride on little twigs that look too slender to support a bird.

ARONIA PERFUME Beside the path by Milburn Pond, I lean down to smell the massed whitish flowers on a bush of aronia. I catch the scent of the blossoms and suddenly a wave of nostalgia sweeps over me. It is nostalgia for some former time, for some former place, where or when I cannot recall. Experienced perhaps long ago in early childhood, the wild perfume stirred into life emotions too deep for conscious thought and too remote for memory.

MAY · 18

BIRD NESTS The female redwing seemed to be stripping bark from a floating stick when first we discovered her. She would alight, balance herself, begin to tug, strip away fibers and then fly off. In our glasses, we saw she was working at a cattail stem floating in the water, swollen and soaked and easily pulled apart. The fibers thus obtained were going into the building of her nest.

In these days of spring, materials, infinitely varied, are being employed in the construction of birds' nests. The time-honored twigs and hairs and rootlets, mosses and feathers, are being supplemented with a surprising number of modern odds and ends.

Not far from here, a wood thrush incorporated an empty Camel cigarette package in the side of her nest. Another wood thrush made use of torn-up bus tickets and a third, nesting near a refreshment stand in an Indiana park, collected discarded pop bottle straws. At Darby, Pennsylvania, a starling added a one dollar bill to its nest and, in a number of instances, English sparrows have made use of cigarette stubs. Small nails, carried from a building project to a birdhouse, formed the steel nest of a house wren.

Last year, at Nassau Point on Long Island, a redstart constructed its nest entirely of insulating material being used in the construction of a new house. At the same place, a prairie warbler made a nest in a bayberry bush from calking cotton used in the seams of boats. Ordinarily, the little thimble nest of a ruby-throated humming bird is coated with bits of lichen. A few years ago, however, I was shown one from the pine barrens of New Jersey that was covered, not with lichen, but with the bud scales of the American elm.

Two miles to the east of me, last spring, a woman hung a frayed, brightly colored blanket on the line in her backyard. A few hours later, she found it more frayed than ever. Baltimore orioles had discovered it and had tugged out threads for the making of their nests. At Palmerton, Pennsylvania, a correspondent writes, a robin flew off with a blue ribbon. Half was cemented into the side of the mud cup of the nest and the other half allowed to hang down the front as though the nest were a blue ribbon winner at a county fair.

MAY · 19

THE JERSEY SHORE With Raymond Bond, I drive down the New Jersey coast to a rambling old wooden inn dating from shorebird-hunting days. The manuscript of a new book rides with us. For two days, ours will be the perfect editorial method. Early and late, we will be out along the shore, where the spring migration of knot and curlew and dowitcher and turnstone is at its height. In the middle of the day we will work on the manuscript—a thin glacier of typed sheets gradually spreading out over the bed, the bureau, the chairs around us.

Toward sunset, today, we drive to Little Egg Point, winding over inlets and across miles of flat sea meadows. Around us was the clamor of the laughing gulls, the soft whistle of the black-bellied plover and the strange staccato calling of the clapper rails. Black cormorants passed in tail-heavy flight, as though they were laboring along with the brakes half on. Once a duck hawk swept low across the salt meadows and all around us, where they had been unseen in the grass before, hundreds of shorebirds—sandpipers, knots, plover, dowitchers, curlew—poured into the air to swirl in tight-knit flocks that form their ancient protection against the peregrine. In the wake of the hawk, the birds dropped down and the meadows seemed to swallow them up again.

Wherever we walked amid the waves of cordgrass, we kept our eyes alert for the bird we wanted most to see, the elusive, sparrow-sized black rail. Under a mounted specimen at the American Museum of Natural History, there is the notation: "Probably the most difficult bird in North America to observe in life." Arthur Cleveland Bent, in his monumental life histories of North American Birds, writes: "Though I have explored many miles of marshes and spent many hours in the search, I have never seen a trace of this elusive little bird." It was, up to that time, the only Eastern bird Roger Tory Peterson had never seen. We, also, failed to see "a trace of the elusive little bird." But our disappointment was expected.

The tide was down when we reached the shore. All over the exposed sand bars and the glistening mud flats, birds were feeding. Least and semipalmated sandpipers ran like feathered mice along the shore. In starts and stops and sudden rushes, the ring-necks—the semipalmated plover—fed. Black-bellied plover and rosy-breasted knots, hudsonian curlews with striped heads, ruddy turnstones in calico plumage, and dowitches exposing white V's on their backs when they took wing—all were in the most beautiful of their plumages, the breeding attire of Spring. None was more handsome than the dunlins, the red-backed sandpipers. Wherever we looked there were hundreds of shorebirds. Continually, flocks rose, turned, alighted again. The light was fading when we drove back, mile after mile, across the salt meadows. Against the glow of the sky, great blue herons were winging their way to their fishing grounds.

MAY · 20

THE BLACK RAIL From now on, the twentieth of May will be known in my own private little calendar as the Black Rail Day. At long last, after so many failures, we saw this day the black rail. We not only saw it but we saw it under circumstances incomparably fine.

Work on the manuscript over, we set out for Beach Haven on

this misty, mizzling afternoon. At the south end of an ocean beach twenty miles long, we came to a low-lying point of land, an area of salt flats and sand dunes, of salicornia and high-tide bushes. A flooded road wound across it with stretches of glistening mud and with, here and there, larger puddles where least and semipalmated and spotted sandpipers fed. Seaside and sharp-tailed sparrows fluttered from the spartina as we advanced. All around us in the mist, we could hear their buzzing songs. From Beach Haven Inlet, ahead, and Little Egg Harbor, to the right, came the clamor of common terns and laughing gulls.

Along this mud road we wandered for half an hour, the smell of salt water, of wet grass, of the sea flats around us, filling the misty air. We had moved to one side to get a better view of a spotted sandpiper when a dark little bird—appearing at first like a fledgling blackbird just out of the nest—fluttered weakly away over the tangled cordgrass. Twenty feet off, it dropped into the grass again and disappeared. As we converged on the spot, a sharptailed sparrow flew up and away. Was this the bird we had seen? We walked slowly ahead, about a dozen feet apart. Our bird flushed again. This time, it crossed the mud road and landed amid sparse grass and salicornia in an open space. Uncertainty was gone. The creature we were viewing in our field glasses was that "elusive little bird," our long-sought rail.

It was only about twice as long as my little finger. From black bill to uptilted tail, its length was hardly that of an English sparrow. The eye with which it watched us was a brilliant, beautiful red. Although of a different shade, it seemed more brilliant, more beautiful than the red eye of the black-crowned night heron. We were seeing the bird in its finest plumage. It was far from a mere *black* rail. A brown wash ran back along its neck and its back was speckled with whitish dots.

It showed little alarm. Stalking it slowly from opposite sides, we got so close field glasses were unnecessary. For nearly ten minutes, we had it within half a dozen feet of us. It would watch us for a minute at a time while we stood motionless. Then, with its neck outstretched, its perky little tail cocked at an angle, its oversized black feet lifting and falling, it would slip through the grass almost without effort. After running thus for a dozen steps, it would stop and watch us again. When it moved, we kept pace; when it stopped, we stopped, too. In this manner, moving ahead, stopping, advancing again, we feasted our eyes minute after minute. It reached the edge of the road and moved out on the glistening mud where it was exposed completely. It appeared slimmer, less dumpy than it is shown in the conventional drawing.

Standing there, it seemed too small, too slight, too delicate to survive in its harsh surroundings. And when it took flight again, its progress was so weak, its journey so quickly ended, that it seemed impossible that it could make long migrations through the air. We could understand the misapprehension of the early settlers who maintained that rails migrated south on foot.

Hardly had our rail disappeared when a chill shower drove us to the lee of a sand dune. We sat in a little hollow among stiff beach grass and watched piping plovers and ruddy turnstones feeding amid the falling rain on a muddy stretch below us. The pelting drops became colder. But we were warm inside, warmed by the finest of fires, flushed with success and aglow with content. We had seen the black rail.

MAY · 21

MAPLE KEYS The gray and white cat comes running to meet me these mornings over a walk speckled with green maple keys. The Norway maples along the way are dropping the small keys with unmaturing seeds. They were stripped off by rains in the night. The keys with growing seeds are still expanding on the branches overhead. In nothing is the prodigal bounty of nature more apparent than in the inexhaustible surplus of flowers and seeds.

MAY · 22

LADY SLIPPERS Naturalists should receive a kind of special dispensation, relieving them of indoor work of every kind, during the wonderful month of May. There is never an end of things to see. Today, it was pink lady slippers.

I drove to northern New Jersey where Ernst Mayr showed me a secret pine wood where the tinted orchids extended across the needle-carpeted floor in dense stands and patches such as I had never seen before. Nodding at the tops of their tall stems, lighted by the sunshine in the open woods, the veined pink flowers were massed in wild gardens, some a hundred yards across. In one place, I counted 84 blooms in a space but two paces square. If all the scattered patches had been put together they must have made a solid stand of half an acre or more of blooming moccasin flowers. For hours I wandered among the pines enjoying this concentration of orchids unparalleled in my experience. Warblers sang, the sunshine was bright, and all around me were the hundreds and hundreds of pink blooms in this secret and magic place. Few people, as yet, know of this isolated tract. All who wish the orchids well will hope for the success of efforts to make this moccasin flower wood a permanent sanctuary.

MAY • 23

UNDER A LEADEN SKY Rain in the night. A heavy, humid morning. Swifts crackling overhead and a veery singing its watery, misty, moisty song in a tangle of trees and bushes.

MAY • 24

SUNFISH NESTS All along the shallow eastern edge of Milburn Pond, in recent days, the sunfish have been scraping away the silt to provide bare, clean patches of gravel for spawning. These patches are from one to three feet across and they extend in an irregular band or chain parallel to the shore and from two to eight feet out from the water's edge. I counted 211 patches this morning. A least sandpiper ran before us as Nellie and I walked along the shore and wisps of mist drifted in parallel lines over the lake. Between these vapor bars we watched the events in the shallow water.

Over each gravel patch was a guardian fish. It rushed toward each interloper and at times a roving outsider would be chased down a whole line of nests, each guardian attacking it in turn so its progress was a series of sharp spurts through the water. Where two gravel patches overlapped, the guardian fish kept rushing back and forth. The fury of the attacker waned quickly as it advanced into the defender's territory while its courage seemed to mount when it was pursued into its own. Thus the seesaw battle continued as long as we remained at the pond. In deeper water, half a dozen feet or so from shore, we could see dimly a whole flotilla of other fish hanging motionless and all facing the nests. The competition was greatest around the larger nests. Over the largest patch of gravel we saw four fish at the same time. The poorer spots and more isolated nests were relatively ignored. The smallest nest we saw was one cleared beside an ancient bed spring, so long in the water that every coil was covered with fine, green, hairlike algae.

All the defenders we saw were males. They build the nests, guard them, fertilize the eggs that are laid among the pebbles of the gravel, often by several females, and defend the young that hatch there. The scoured patches in the shallows were a sign that the pond water had risen to a temperature of 68 degrees F. It is at that point, each spring, that the sunfish, the bream, the pumpkin seed, begins the making of his spawning nest along the edges of the pond.

MAY · 25

NEMATODES AND UPLAND PLOVER Several years ago, golden nematodes appeared in the potato fields of Long Island, resulting in a quarantine and a series of years when the fields remained unplanted. As a result, I was told today, upland plover increased in the area during this temporary expansion of fallow land. Nature fills every chink. No vacancy is left vacant for long.

MAY • 26

GRASS IN SWALLOW'S NEST Five days ago, on the 21st, I noticed the first dark, damp mud of a barn swallow's nest on the side of one of the roof beams of the wagon shed at the top of my Insect Garden. Since then, the mud cup has taken shape and is now completed. Watching the birds come and go in their building labors, I realize as never before how much grass goes into the construction of a barn swallow's nest. It is far from a dwelling made of bricks without straw. Now, from the finished nest, a kind of dangling beard of blades and stems hangs down. I count the grasses in this beard and find they number more than fifty. The longest is nearly a foot in length. In this particular nest, there seems an abnormal amount of grass. But when I examine an old mud cup on a rear beam, a nest with no grass visible on the outside, I find beneath the outer mud a feltlike material that reveals the quantities of grass that went into its construction. Later in the summer, when the swallow brood is gone, I plan to take down this nest I have seen constructed and analyze its content. What else is in it besides mud and grass? Ask me again in September and I will know!

EGRET A white bird in the sunset—the first American egret we have ever seen flying over the yard—passes above us as we look up about six o'clock this evening. This brings the list of birds seen in or from the yard to about 100.

MAY • 27

DAYBREAK Wandering over my Insect Garden hillside, a few minutes after five this morning, I hear the "K-hick! K-hick! K-kick!" of a Virginia rail coming from the swamp. An oriole's rich and rounded song descends from one of the apple trees and high in the swampside maples, touched by the sun, redwings call. Along the packed earth of the trail I follow, little disks of darker earth mark the ant hills newly made in the night. Here, too, I meet a late home-going garden slug crossing the dew-wet path. All of the mitten leaves of the sassafras, all of the sword leaves of the cattails, are spangled with droplets of dew. Out in the lowland, the blue of wild iris is now scattered among the green leaves of the sweet flags. A long-billed marsh wren ascends above the cattails on quivering wings. It bursts into its first clattering little song of the morning. Day has begun and I go home to breakfast.

MAY · 28

WILD CHERRIES Now the wild cherries are in bloom. Wherever I walk, along the swamp, on my garden hillside, beside Milburn Pond, the white foam of the massed florets shines against the dark and varnished leaves. Small wild bees hover and alight, harvesting minute quantities of nectar and pollen. Viewed from a distance, some of the trees are so covered with airy masses of white blooms that they suggest cumulus clouds billowing into the air. I count the components of half a dozen masses and find that there may be half a hundred florets in a single mass. How many tens of thousands of florets the massed blooms of a tree contain, I can only guess.

MAY · 29

CUCKOOS AND CATERPILLARS A few years ago, an abnormally warm spring, followed by a sudden cold spell, almost wiped out the tent caterpillar population of this region. Previously, yellow-billed cuckoos nested in the apple trees. In six minutes, a yellow-billed cuckoo was seen to consume forty-seven tent caterpillars. So fond of these larvae are the birds, the hairy caterpillars make up so large a part of their diet, that the stomachs of cuckoos are sometimes covered with a thick lining of hairs. In recent years, with the tent caterpillars gone from the wild cherry trees, the mellow "kow-kow-kow-kow" of the rain crow has also been absent from the orchard hillside. Records all over southern and central New York have indicated a scarcity of cuckoos. But, when a widespread infestation of tent caterpillars occurred in Ontario during a recent summer, cuckoos were numerous. Like the snowy owls of the north that move back and forth across the Arctic barrens according to the abundance or scarcity of lemmings, the cuckoos may concentrate in the areas where their smaller larval game is most abundant. Last spring, a few tent caterpillars spun their silken webs in wild cherries beside the swamp. This spring, I count a dozen of their tents in the space of a hundred yards among the higher trees. And, in the warmth of noon today, I hear once more the welcome "kow-kow-kow-kow" of a cuckoo returned again to my garden hillside.

MAY · 30

THE NESTING OF REDWINGS Over the length of the swamp, these days, the male redwings mount like intercepter planes to harry every larger bird that passes by. It is nesting time in the cattails and the blackbirds attack harmless bittern and marauding crow alike.

Thirty feet separate the highest and the lowest redwing nests I have seen this year. Among grapevines, at the top of a thirty-foot tupelo tree near Milburn Pond, one pair has built a nest. In the tangled grass, far out on the sea meadow, another pair has produced a nest that touches the ground and has a mud bottom. A few seasons ago, a peculiar redwing nest made its appearance close to the edge of the sea meadow. It was built, rather early in spring, in a clump of fiddlehead ferns. The ferns grew irregularly so one side was lifted higher than the other and the nest tilted at a steep angle. In spite of this, the eggs were laid and the redwing brood successfully fledged.

LITTLE EARS As I crossed the hillside this morning, a small patch of dry, yellow grass, beside a green clump, caught my eye. Carefully, I pulled aside the grass and came to a soft, gray blanket of fur. Parting this, I exposed the little ears of a nest full of baby cottontails. Just as carefully, I replaced the fur coverlet and the covering of grass. In a little while, now, I will see new rabbits hopping about my hillside.

MAY · 31

ANTS AND APHIDES I stop, this morning, beside the clumps of viburnum where the hillside path reaches the edge of the swamp. Scores of the leaves are curled downward and on the underside, as though sheltered within a green barn, are masses of close-packed little plant lice. The aphides, in their sap sucking, have caused the leaves to curl. And, here, black carpenter ants are standing guard among their herds and gathering the sweet honeydew given off by the aphides. Milkmaids at work at dawn!

I tapped a stem and all the black ants for a foot around began running helter-skelter, this way and that, in search of an enemy. Nothing disturbed the aphides. I leaned close and breathed on the nearest cluster. The guardian ants went wild but the aphides were unmoved; they continued their sap drinking.

Four of the aphid clusters had in their midst the deadly carnivorous larvae of the fly, *Syrphus americana*. Although these larvae spend their days devouring the very herds the ants were guarding so zealously, the ants paid no attention to them. I see them pass over them time and again, ignoring them, apparently unaware of their destructive activity.

THE LUNA MOTH The delicate beauty of that most ethereally beautiful of all the American moths, the Luna, unfolded before us today. It emerged from a cocoon I had brought indoors in the autumn and clung motionless while its wings and long ribbon tails expanded and hardened. We put it on an iris. It climbed to the top. There it remained for hours, quiescent, awaiting the coming of night.

CHAPTER SIX

June

JUNE • 1

THE DAWN OF JUNE To the Insect Garden, a little after 5 A.M., on this first morning of June.

The machinery of nature, with its winds and dews and dawns and morning mists, produces poetry as well as seasons and growth and change. Here the functioning of nature's cogs has created dew drops and veils of luminous mist caught among the cattails. Before the work of the day, taste the poetry of the day! Our poor, battered minds and spirits need the dawn. There is the calm of nature, the sanity of the earth, in each breath of scented air on a sunrise in June.

In the dawn calm, those Lilliputian argosies, the woolly aphides, are setting sail over the clumps of wild cherries. A robber fly clings to an apple leaf, awaiting the passing of its prey. First the gnats, then the robber fly; first the prey, then the predator—this is the invariable sequence of returning life in the spring.

JUNE • 2

VOICE IN THE DARK Night sounds are mysterious. Often they are eerie and fearsome. The call of an owl in the dark woods, the croaking of a night heron in the black sky, seem to reach us from awesome and foreign realms. But on these latter days of spring, when darkness falls, one voice comes from the dusk that has a friendly and familiar ring. This is the call of the whippoorwill. There is nothing disturbing or alarming here—perhaps because the bird talks our language. At any rate, when I was a boy in the Indiana dune country, returning home late with the cows, this repetition coming endlessly from the dark woods and dimly lighted pasturelands always was somehow calming and reassuring.

A thousand miles, and many years, separated those calls from the ones we heard tonight when we drove out on the island to listen to the whippoorwills. Yet the effect, in spite of time and distance, was much the same.

Everyone is familiar with this nocturnal song. Yet perhaps not one-tenth of one percent of those who have heard the call have seen the bird that produces it. Up until the early part of the last century, nobody knew that the whippoorwill was its source. Until Alexander Wilson demonstrated differently in 1831, the voice of the whippoorwill was thought to belong to the nighthawk. Before that, these two birds of the twilight were believed to be one and the same.

The migration which brought the whippoorwills back this spring, as has been the case in countless springs before, consumed thirty-five or forty days. Both in the speed of travel and in the time of year when the journey is made, the migration of the whippoorwill and that of the chimney swift coincide. Both depend upon insects for their food.

Even those great moths of the spring night, the Polyphemus, the Cecropia, and the pale green, ribbon-tailed Luna, fall victim to the whippoorwill. One naturalist, who watched a whippoorwill take off from a branch of an oak tree, noticed that it opened its mouth *before* it launched itself into the air. As many as thirty-six full grown moths have been taken from the stomach of a single bird. In addition to the larger nocturnal insects, it consumes immense numbers of mosquitoes.

Direct the beam of a flashlight toward a resting whippoorwill and its eyes will glow a brilliant orange when struck by the rays. John Burroughs once counted the calls of one whippoorwill and found that it repeated its song 1088 times with hardly a pause for breath. In early spring, the voice of the twilight is the calling of the marsh peepers; in late summer, it is the strident shrilling of the insect orchestra. But now, in these June evenings, the voice of the twilight is the song of the whippoorwill.

JUNE · 3

COWS AS BOTANISTS As I walk down the hillside of the Insect Garden this morning, and notice the patches of plantain in bloom, I am reminded of the wisdom of the farmyard. Seventy years ago, Asa Gray, America's greatest plant authority, paid tribute to the cow as a botanist. There are, he wrote in a paper in the *American Journal of Science*, two closely related kinds of plantain in the United States that had been separated only after long study by trained botanists. But the cows had known they were different all the time. They ate one kind of plantain but passed the other by.

THE EGG CASE OF THE MANTIS Last winter, I brought home several twigs to which were attached the walnut-sized, tan-colored egg cases, or oöthecas, of the praying mantis. One was blackened on the outside, charred by flames from a grass fire. I had little idea it would still contain uninjured eggs. Yet this morning, a little before nine, the small, soft, honey-yellow insects began to emerge. They soon festooned the twig like a living cobweb; more than 100 hatched and emerged during the morning.

JUNE · 4

THE SQUIRREL'S CRUTCH One of the gray squirrels that shares our yard with the redoubtable Chippy has injured its left hind leg, perhaps in a fall from a tree. It gets about on three legs but, when I throw it a peanut, it is unable to sit up, squirrel fashion, to eat it. Several times, in recent days, we have seen it do the same thing. A forked branch, one of the limbs lost by the dying maple, has been left at the edge of the rose garden. The crippled squirrel carries its nut to this branch and braces itself against it, sometimes in the crotch, sometimes on the side. Thus supported, it eats its nut sitting up in the traditional manner of squirrels.

JUNE • 5

PAINTED TURTLES Under the heat of the midday sky, four painted turtles are sunning themselves on a log that juts out into the swamp stream below the Insect Garden. One is full grown, the others are baby turtles only three or four inches long. For ten minutes, as I watch, a kind of game goes on. The three young turtles seem to be playing cock-of-the-rock with the top of the adult's back as the rock. First one and then another will clamber up the slippery shell to maintain its position at the top until displaced by another climber. When two are near the top, they push and shove until one loses its hold, slides down the slope of the shell and splashes into the water. All the while, the adult turtle basks placidly in the sun.

It seems asleep. But it is not. I move among the sweet flags to get a better view. Its eye catches the movement. In an instant, it has tumbled off the log into the water, carrying the current cock-of-the-rock with it. The other baby turtles scramble over the edge of the log and disappear in the brown water. I have noticed, in watching the painted turtles of the swamp stream, that the older ones are always more wary. They are the first off a log; the little turtles are the last to leave.

Minutes go by. Then the nose of the adult turtle bobs up, like a small, dark cork, a few yards from the log. Around it, three tiny turtle noses break the surface and the quartet swims leisurely back to the log. Within five minutes, all is as it was before; the adult basking in the sun and the three baby turtles contesting for the highest position on its shell—pushing, scrambling, slipping, splashing into the water and returning to the log once more.

JUNE · 6

WILDLIFE W. H. Hudson, one of the greatest of literary naturalists, used to express his attitude toward wild creatures with the words: "Neither pet nor persecute." His outlook was sound. The wild should be left wild. Birds and animals should be permitted to follow their instincts and live their natural lives. I have no desire to persecute; but I find it harder not to pet.

JUNE · 7

SEA LETTUCE In the long, slow fading of the light, this evening, Nellie and I sit on a cushion of dried sea lettuce at the edge of the bay and eat our supper of sandwiches and tomatoes. Tossed ashore by successive storms, the seaweed is folded into parallel lines or corrugations. The bay appears to have a pleated hem. This wrack of the sea lettuce is bleached gray on the shore; is still green where it touches the water.

Spotted sandpipers and sanderlings feed on the wrinkled plain of cast-up seaweed. A least tern hangs in the wind, dives, flutters in midair, continues its fishing close by. Once a "cut-water," a black skimmer, with deliberate beats of its long and slender wings, comes low across the bay to fish in the inlets. It follows their windings and swings from tidal creek to tidal creek across the salt meadows with its red lacquered bill almost touching the vegetation. It is a "cut-grass" as well as a "cut-water."

Walking home we noticed how we can stimulate the clapper rails into action by clapping our hands together at the usual tempo of their calling. Seaside and sharptailed sparrows sang from the swirled crests of the cordgrass. The song of the former is the more robust. The little sharptail has a thinner song and gives the impression it has a slight cold in its head. Both birds have modest, unassuming little songs, as unpretentious as the fluttering flights they made close above the grass clumps. The shortest bird song I know is that of another inhabitant of the salt meadows, the Henslow's sparrow. As I heard it near Tuckerton, on the New Jersey coast, the sum total of its vocal effort is a succinct "Flea-Sick!" It is all over in about one-fifth of a second.

JUNE · 8

REDWING FORAYS During recent days, I have been noting the birds that redwings pursue through the sky above Milburn Swamp. They are eternally on guard and challenge any airborne creature, large or small, that passes above their nesting grounds. I have seen them rise to attack and harry birds as large as the American bittern, the green heron, the black-crowned night heron and the herring gull. At other times, they have pursued a slate-blue pigeon, a fluttering least bittern, several crows, a grackle, barn swallows and tree swallows, a killdeer and a greater yellowlegs. I even watched one doughty redwing, in a spurt of wild gyrations, pursue a speeding chimney swift.

JUNE · 9

CHANGES OF THE SEASON With consequences innumerable, the season advances day by day. Sunflower sprouts, the hull of the seed like a clothespin on one of the twin leaves, have risen from the ground at the Insect Garden. Green aphides have appeared on the new stems of the rambler roses. On the hillside, there is a sprinkling of white daisies and, beside the swamp, the demure flowers of the blue-eyed grass look out from the lush vegetation. Ragweed is rank where new dirt was piled from an excavation. Migration is over. Blackberry tangles already are white with flowers. Cygnets are in the swan's nest at Milburn Pond. The Baltimore oriole has been singing only snatches of his song and fledglings of various kinds are out of the nest, following their parents about, fluttering their wings and begging for food. Each flower and creature returns in its season and June is the season of many things.

JUNE · 10

SWEET FLAGS The path down the Insect Garden hillside winds past a wild cherry tangle, between two scraggly clumps of arrowwood, *Viburnum dentatum*, and out among the sweet flags that mass their green at the edge of the swamp stream. This morning, all the leaves of the hillside are tossing in the wind, the broad leaves of the maples, the glistening, varnished spearheads of the wild cherries, the green hearts of the lilac bushes. Whipped by the gusts, the frosty-gray undersides of the buddleia leaves appear and disappear. And, all around me, at the streamside, the slender, iris-like leaves of the sweet flags swing and clash as though hundreds of swords were wielded in battle.

Each spring, these leaves shoot upward from the interlacing mat of knobby roots in the wet soil below, shoot up as though they would reach the sky. For them, the beginning and the end of growth is early and the summer is a kind of long decline.

I do not know how these sweet flags got their start in this stretch of marshland. But I do know that early pioneers carried the roots with them as they spread westward. They valued them highly as a cure for ague, dyspepsia and an assorted list of ills. They planted them in swamps and around water holes to provide themselves with a medicine chest out-of-doors.

In those early days, the uses of the sweet flag roots were many. As a powder, they appeared in toilet mixtures and in snuff. Oil, extracted from them, went into the manufacture of perfumes. Frontier housewives ground them up and used them as a flavoring substitute for nutmeg and ginger. As a candied sweetmeat, they were regularly sold on the streets of Boston. And, in Europe, pieces of the root were chewed by boys as a sort of prechicle chewing gum.

The sweet flag, *Acorus calamus*, has many common names. It is known as flagroot, myrtle grass, myrtle sedge, gladdon, cinnamon sedge, sweet sedge, sweet rush, sweet cane, sweet root, sweet myrtle, myrtle flag, beewort and sea sedge. It is now found from Maine to Minnesota and as far south as Texas.

At first glance, its slender leaves seem to link it with the iris family. In truth, however, it is a close relative of the skunk cabbage, the golden club and the Jack-in-the-pulpit. Four centuries ago, it was the leaf of the sweet flag that contributed to the downfall of Cardinal Wolsey in the days of Henry VIII. He had such leaves strewn daily over his floors to perfume the air. They were brought at considerable expense from the marshes of Norfolk and Suffolk. Among the charges that led to his tragic last days, as recorded in William Shakespeare's *Henry VIII*, was the one that he had been extravagant in his spending for sweet flag leaves.

JUNE • 11

WONDER Out-of-doors, adventures are everywhere. Wonders are all around us. If the world is stale, its fascination gone, the fault, we find, is in ourselves. As G.K. Chesterton put it: "The world will never starve for want of wonders, but only for want of wonder."

JUNE • 12

KING CRABS Again to the sea moor at sunset. For a time, I stand beside a little scollop of the bay, edged with a windrow of dry seaweed, sunbleached and winding among the wiry grasses like the sloughed skin of a serpent. An alarmed redwing flutters overhead, repeating again and again his call of concern: "Check! Check! Eeeeek!"—the ending as shrill as the squeaking of a hinge. Seaside sparrows buzz and, away across the meadows, clapper rails, from time to time, begin calling like puffing little locomotives getting underway.

I am only partly aware of the sounds around me. For I am again a spectator at one of the most ancient dramas of the earth. Once more, as during ages immemorial, the king crabs, those oldest dwellers in the sea, are coming to the shallows to fertilize and leave their eggs behind.

Moment by moment, the water creeps ahead as the tide runs in. The crabs, the "horsefoots" of the baymen, are shadowy, then distinct, as they appear from the murky water. They come linked together, the smaller male behind. Occasionally, a dark, horseshoe-shaped shell appears above the water to sink again with lines of foam and sediment streaming back along the surface. Farther and farther, they push up into the shallows to deposit the translucent little globes of their eggs. This is the great annual event in the lives of these creatures that have lived on and on after all their early contemporaries have become fossils. These events, as I watch them in the twilight of this June day, are the same as they were a hundred million years before the dinosaurs. In an unbroken chain they link the Atomic Age with the primitive world as it was 200,000,000 springs before man first walked the earth.

JUNE · 13

CAN DRAGONFLIES FLY BACKWARD? Four and twenty blackbirds—exactly the number in the famed pie of the nursery rhyme—were congregated in the top of the willow tree beside the swamp stream when I came down the trail this afternoon. They scattered across the cattails and I pushed my way through the sweet flags to the edge of the stream to watch the dragonflies. This summer I hope to find the answer to a question of years' standing. Dr. Frank E. Lutz, then head of the Department of Insects and Spiders at the American Museum of Natural History, some years ago, checked over the manuscript of one of my early books. In the margin, beside the sentence: "Dragonflies can also fly backward," he wrote: "Can they? I used to think so but now I wonder." Ever since, I have wondered, too. A score of times, I have seen dragonflies move backward. But in every case I was not able to eliminate all other possibilities, such as that they might be drifting with a breeze. This afternoon is perfectly calm. But all the dragonflies I see are moving ahead or standing still, hanging over the mirror of the brown water on fluttering wings.

JUNE · 14

THE PLEASURES OF INSTINCT Who can doubt that it "feels good" to the turtle to sun itself on a log; that it "feels good" to the flicker to rap with its bill on hollow wood; that it "feels good" to the muskrat to dive into water? Pleasure and pain, comfort and discomfort, these are the push and pull of the motor of instinct. There probably are as many different kinds of pleasures as there are different kinds of instinctive acts.

JUNE • 15

THE GUARDIAN BIRDS About four o'clock this afternoon, I discovered a running attack going on in a maple tree. A gray squirrel was leaping from limb to limb, seeking to escape from a diving bluejay. Whichever way it turned, the bird was close behind. It finally found sanctuary near the trunk, secreting itself where two limbs projected out, one close above the other.

This bluejay, for the past day or so, has been on the aggressive, diving at cats, chasing squirrels, driving away other birds. The explanation? Its fledglings are almost ready to leave the nest. I discovered this nest by chance in the thickest part of a spruce tree. For bluejays, unlike robins and most other birds that reveal the presence of their nest by mounting excitement when you draw near, remain perfectly silent no matter how close you come. If you discover their secret, you do it without any help from them.

Newspapers, during the latter days of May and early June, are likely to carry stories about mystery birds diving on pedestrians in unprovoked attacks. Three such items have come to my attention, each originating in a different part of the country.

1. Pedestrians walking along a tree-lined New England street are being attacked in the dark. Vicious birds dive out of the shadows. They sweep close. They pursue the innocent walker down the street, diving again and again until he has outdistanced them.

2. Bluejays in a midwestern town seem suddenly to have gone mad. They begin plunging like dive bombers on everyone passing beneath certain trees. One man has had his hat knocked off. Several women have been badly frightened. Even dogs and cats passing by are victims of the attacks.

3. In another American city, the police are called to shoot "mystery birds" that are "terrorizing" a certain suburban area after dark. Children are being kept indoors because of the attacks and people returning home late are giving the area a wide berth.

Almost always, these bird bombers are screech owls or bluejays. And, invariably, their attacks begin just about the time that their young, in some near-by nest, are approaching the time of their first flight. The brave parent birds are simply trying to protect their young by driving away everything that represents danger to them. Too often, the conclusion of the story has been that the police have been called in and the "vicious birds" have been shot. A little understanding of what really is happening would prevent such needless tragedies.

The birds are neither vicious nor attacking without cause. They confine their activities to a small area near the nest. And their attacks last only a few days. Even then, they are largely games of bluff; the birds merely dive close and veer away. Frightened pedestrians who demand that the police shoot the "dangerous and vicious birds" rarely realize that this procedure is likely to leave helpless young to starve.

JUNE • 16

A SNAPPING TURTLE LAYS HER EGGS In open sand, close to the Insect Garden and a hundred feet from the edge of Milburn Swamp, this morning, I came upon a snapping turtle laying her eggs. She had pushed the front edge of her shell into the soft dirt and gouged out a trench two and a half feet long at the end of which she had excavated a pit with her hind feet. In this hole she was depositing her spherical eggs. As each rolled down the slope into the pit, the turtle lowered a hind leg and, reversing the foot, moved or shunted the egg into place. The final three eggs were laid in five minutes. Then the turtle rested a long time. Finally, with slow movements of her hind feet, scraping the sand ahead alternately, she filled in the hole. I first discovered her about seven o'clock in the morning. It was ten o'clock, three hours later, when her work was done. Then the snapper, about a foot across and with the green "moss" of algae along the rear of her shell, started for the water. She moved ponderously and never once looked back at me or at the spot where her children one day would appear as though by magic from the ground.

TURNABOUT TURTLE This day has been a turtle day. First the snapper, then a painted turtle I came upon crossing the same level stretch of sand. It was heading east. I picked it up, turned it around and set it down, facing the opposite direction. It made a half turn and plodded east again. I tried the same experiment over and over. Although the ground was level and no slope present to give it a clue, the turtle always swung like a compass needle toward the direction in which it had originally been heading. Twenty-two times I reversed its position; twenty-two times it turned back again. In the end, I turned it on its back, spun it rapidly a dozen or more revolutions and then set it on its feet heading in the wrong direction. It was not in the least confused but swung again to the east. This time I let the harassed turtle proceed unmolested along the sure path of its determined course.

JUNE · 17

SCHOOL I always want to be a schoolboy on the last day of school!

JUNE · 18

THUNDER PUMPER Beyond the sluggish water of the brown swamp stream, the forest of cattails rises in a dense wall of vertical lines. On the edge of the stream, before the cattails, frozen in one position, its bill pointing skyward, its body merging amazingly with its background, stands an American bittern, the thunder pumper of the swamp.

No other member of the heron tribe is more shy, more secretive, than the bittern. Neither of the two great pioneers of ornithology, Alexander Wilson and John James Audubon, ever saw a bittern's nest. Even the eggs of these elusive swamp dwellers are camouflaged. They are almost exactly the color of the dead flag leaves of which the nest is usually formed.

It has been reported by competent observers that when an intermittent breeze blows over a stretch of marshland, an upright bittern will remain motionless as long as the air is calm. But when the cattails around it begin to wave in the breeze, it, too, sometimes begins swaying. Thus its form and vertical stripings help make it inconspicuous under either condition.

It is largely through the love song of the male, one of the most strange and unmusical serenades in the world, that we know the American bittern. Its sucking, pumping, booming sound will carry as far as half a mile across a quiet marsh. I have heard it for weeks, now, most often at dawn, and twilight. The sound has been recorded in such phrases as "pump-er-lunk" and "plum-pudd'n." It has been likened to the noise of an old-fashioned wooden pump and to the sound of a stake being pulled from soft mud. It has given the bird its common names of stake driver and thunder pumper. On at least one occasion, a bittern has been heard pumping thirty feet up in a tree.

Almost every movement of the bittern is stealthy and deliberate. Its feet are lifted so slowly the movement is almost imperceptible. They are set down with equal care. Thus it draws close to its prey. Frogs, fish, small snakes, meadow mice, crayfish, grasshoppers and even dragonflies are speared by the lance bill of the bittern.

During the early days of this month, the female has begun incubating the eggs in carefully secreted nests among the cattails. This continues for twenty-eight days. The baby bitterns, from three to seven in number, remain in the nest for an additional two weeks after hatching. Their first plummage is buffy down that, like the eggs, matches the color of the dead flag leaves around them. Later, their bodies gain the vertical striping which plays such an important part in their secretive, adult way of life.

JUNE · 19

FROGHOPPER TIME Glistening globes of white, each about the size of a pea, shine out from the grass tangles of the garden hillside this morning. I count half a hundred along the sloping path to the swamp edge. Each is a mass of froth, like the beaten white of an egg, produced by a tiny soft-bodied, immature insect inside. Here the insect lies, moist and hidden, sucking sap from the grass stem, safe within its fortress of bubbles. Commonly known as the froghopper, the little Cercopid uses a mechanism unknown elsewhere in all the world of living animals. For upward of 10,000,000 years, froghoppers have literally been saving their lives by blowing bubbles. I see their little foam castles all summer but most frequently in the spring. Now is froghopper time. When growth within the mass of foam is ended, the bubble-blower develops wings in its final molt, becomes a nondescript brownish little bug, allied to the tree hoppers, and flies away. Rarely is it noticed in its adult form. Its great achievement, its claim to fame, is this shining house of foam that is produced during its earliest days.

JUNE • 20

WOODPECKER MARKINGS I notice today, for the first time, how the white and black markings on a downy woodpecker give the bird a sudden and striking camouflage when it alights on a tree where lines of lichen run across the bark.

TURTLE AND DRAGONFLY To the Insect Garden about 4 P.M. As I stand motionless among the sweet flags beside the swamp stream, a snapping turtle, perhaps the one I saw laying eggs the other day, plows its way deliberately through clouds of filmy green algae. Once it rests at the surface, its feet trailing, its thick-necked head outthrust. And as it floats there, a smoky-winged dragonfly drops down, alights on its back and drifts with the drifting turtle. It rides thus for a minute or more. Then the turtle begins to swim and the dragonfly zooms aloft and hangs on quivering wings in the sunshine.

Does it move backward as it hangs there? I am positive it does. Then the question is settled. But, no! I remember that, as I stood there, a slight breeze blew and died away and blew again. Can I be sure that the breeze did not move the hovering dragonfly to the rear? I cannot. So I will have to watch another day!

JUNE · 21

A ROBIN DAY On this twenty-first of June, the hinge day of the seasons, the yearly tide of light reaches its flood. Tomorrow, it will begin the long rollback to the dark days of December. I heard robins singing this morning shortly after four o'clock at night. A robin's day is a long one. It uses up all the daylight, even on this longest day of the year.

THE MEADOW TWILIGHT In the later sunset of this final day of spring, we walk to the bay. All across the salt meadows, sparrows, perched on the waves and swirls of the cordgrass sea, buzz and tinkle their evening songs. The seasides predominate close to the water with a few sharptails; sharptails predominate inland along with a few seasides. A white line that runs along the "lower jaw" of the seaside at times gives the impression it is a whitethroat. Once we saw a sharptail alight with a white moth in its bill and for a moment mistook it for a seaside. Sharptails have pronounced yellow, or ocher, markings on the face. Of the two, the seaside is larger, being about the size of the English sparrow. It chirps at times in a way that suggests the whitethroat at evening.

Close to the bay, we saw one seaside sparrow rise in the sunset to a height of twenty feet or more, fluttering up and up, singing a little sky-song as it rose. It hung there, the edges of its wings translucent in the last rays of the sun so that it seemed surrounded by a luminous cloud of shimmering light. Here it remained suspended for a moment or two, a singing, shining mote of life above the darkening expanse of the meadows.

Two white-gray lines of sea lettuce, left by storms about four feet apart on the land, run like wagon tracks in the grass, following each curve of the shore. As we stand there in the quiet of the evening, we hear the knife-whetting "Wheet! Wheet! Wheet!" of spotted sandpipers at the edge of the water. Least terns flutter and plunge in their minnow fishing, and fiddler crabs appear from their holes and run over the mud below us. A faint mist forms in the air.

And so this longest day in the year comes to an end with silver mist and low-lying land and the smell of the sea. Twilight here is doubly impressive for we are face to face with twin mysteries—the mystery of the sea and the mystery of the night. We, as diurnal creatures of the land, are looking into foreign realms, into worlds other than our own, into the mysterious dark and the mysterious depths.

JUNE • 22

THE BACKWARD BLUEJAY It rained before dawn. I awake to find one of those moist, still mornings when the sky hangs low and sounds carry far through the damp air. One of these sounds is the persistent and raucous calling of a young bluejay. It comes from the top of the spruce tree. It trails away down a line of maples. Then it grows suddenly louder, coming from the cedar just outside my window. This is a sound with which I have become familiar in recent days. This is the sound of the Backward Bluejay.

In almost every brood, one fledgling seems more dependent upon its parents than the others. It is the last to break home ties; the last to learn to feed itself. Through timidity or inertia or sheer laziness, it remains dependent as long as it can. All the other bluejay fledglings have learned to forage for themselves. This one trails after the parents, squawking its demands, clinging to a branch and fluttering its wings as it begs for food. This morning they have been rebuffing it. Each time it seeks a handout, they utter a harsh scream and fly away. It sets out in pursuit, shouting its demands. And so the chase continues from tree to tree. It is a healthy, normal bird, as well able to take care of itself as its brothers and sisters. But being fed is such a pleasure unalloyed by labor that it seems to be determined to enjoy it as long as possible.

I do not know whether it was this same Backward Bluejay or not, but the other morning, we saw a young bird come for a landing on the clothesline. It failed to brake its speed sufficiently, grasped the rope and continued in a half loop forward. It hung upsidedown not knowing what to do for fully a minute. Then it let go, spread its wings as it fell straight down, and curved upward into flight before it struck the ground.

CRITCHICROTCHES Golden thumbs, some nearly three inches long, now thrust out at an upward angle from the sides of special leaves among the sweet flags. Each flower-encrusted spadix resembles a miniature ear of corn. It is in these fleshy floral spikes that the sweet flag shows its relationship to the golden club and the Jack-in-the-pulpit. And, like those plants, it literally runs a fever at blooming time, generating sufficient heat to raise its temperature above that of the surrounding air. As I stand watching a painted turtle explore the opposite stream bank, I nibble at one of the flower spikes. Like the leaves and roots of the sweet flag, the spadices also have a pleasing taste. In New England, such flower thumbs were known as critchicrotches. Henry Thoreau's Concord journal contains such references as "Critchicrotches have been edible some time in some places" and "The critchicrotches are going to seed."

JUNE · 23

ANT LIONS Once again, in open sand among the Hills of Lilliput in my Insect Garden, ant lions have dug their round pits. Secreted at the bottom they are now awaiting, with hollow, pincer jaws, the descent of some unwary victim. Nothing comes their way during all the time I watch the pits this morning—a period comparable to months in the life span of a human being. Hawks and weasels—the active pursuers—have to eat frequently. Ant lions and spiders—the inactive waiters—can go for periods unfed. At one point on the sand, I come upon the body of an ant lion, dead for several days, now partially shriveled. In a kind of retributive justice, a score of tiny red ants are feasting on the dry husk of its flat, seedlike body.

JUNE · 24

SMACKING OF BROWN THRASHERS To the garden in mid-afternoon. Heavy, humid heat presses down on the swamp, the hillside, the Insect Garden. Birds are silent, resting. Only the insects are more active than ever. Heat is the elixir, the tonic, the great restorative of their little lives. Dragonflies dart and zoom and hang suspended over the swamp stream. Carpenter ants, running in high gear, forage up and down the bark of the apple trees. Black prince damsel flies and grayling butterflies flit down the swamp path before me.

I strip off my shirt and, for a time, toast in the sun, stretched out in an open space between the trees. A tiger swallowtail, crossing the Insect Garden, trails its shadow across me as it passes by. Lying there, dozing in the heat, I suddenly hear the silence broken by a loud, metallic smacking beyond the wild cherry tangle. It is repeated over and over, grows louder, comes nearer. Then a gray tomcat, its ears cocked back to catch the sound, walks down the path with a pair of brown thrashers flying from limb to limb behind it, harrying it, calling attention to it, warning all other creatures with their loud smacking call of alarm.

JUNE • 25

FLITTERMICE In the long June twilight tonight I watch the bats, the flittermice, zigzag in their wildly staggering flight above the cattails. How beneficial they are; how unearned and unjust is the long enmity that has surrounded them! I wonder if this enmity will be lessened now that the secret of their uncanny ability in avoiding objects in the dark has been exposed as similar to the sonic depth finder man uses at sea. No longer does the skill of the bat seem allied to sorcery. Understanding has replaced mystification. And this, it is to be hoped, may lighten the load of dislike that has descended from ages when the human mind viewed the natural world through superstition's dark and distorting glass.

JUNE • 26

SQUIRRELS IN A HEAT WAVE The heat of recent days reached its peak today. The mercury stood at above 95 degrees F. on the shady side of the house. All creatures panted in this succession of furnace hours. I panted with them. I also noticed the squirrels in the heat wave. One was stretched out on a limb, its forepaws hanging limply down. Another spent a long time flattened out on the cool bricks of the shaded front walk. It was flatter than I had imagined a squirrel could get.

JUNE · 27

THE GREEN GNATS Beautiful gnats, pale green, almost the color of a Luna moth, drift down the slope of the hill in the quiet air of sundown. Most such insects are gray or nondescript. These are delicate things of beauty, works of art in miniature. As the sun fades, after the glare and heat of the day, a few crickets on the hillside begin chirping. The sound seems like the first tuning up of the great insect orchestra that will be in full swing in late September days. From the marsh comes the low "coo-coo-coo" of a least bittern, a mellow call like that of some faraway cuckoo or muffled owl. And, somewhere among the upper branches of an apple tree, a mourning dove repeats its drowsy call, a soothing sound of indolence as though the bird were nodding, on the verge of slumber. All things seem subdued by the heat of the day except the indefatigable marsh wren that rises above the cattails on its little song flights, as strident as a clattering lawn mower in action.

JUNE · 28

EDIBLE HOMES Imagine yourself living in a globelike room with greenish walls bulging outward and upward and then arching in to meet above your head. Imagine such a room constructed of succulent, edible material, forming a house that at once provides food and shelter, plenty and protection. That is what you would find if you traded places with one of those gall insects that now live in the globular swellings on the stems of my hillside goldenrod.

These swellings are produced by the female fly, *Eurosta solidaginis*. At the same time she lays her egg in the plant tissue, it is believed, she injects the chemicals that cause the swelling. By this simple act, the fly provides its offspring with an edible home where it can live out its grub days free from want and protected from its enemies.

As I walk about my Insect Garden, on these early days of summer, I see a host of other galls, swellings of varied forms on the leaves of trees and on the leaves and stems of plants. While the globular galls of the goldenrods are the product of a fly, other galls are made by moths, by various small wasplike insects and by mites. Each gall-maker specializes in its own plant or tree. These run the gamut from wild lettuce, touch-me-nots and cinquefoil to grapevines and huckleberries and wild cherries and blackberries and rose bushes. They include dogwood and witch hazel and sumac and alder, cottonwoods and conifers and oaks and hickories and willows and elms and hackberries and maples and tupelo trees.

Sometimes the galls are on the stems and twigs, sometimes on the leaves themselves. They resemble seeds and cones and fungi and insect eggs and masses of fuzz and burs and nuts and flowers and cockscombs and fruit. Some are green and some are red and some are shaded with delicate tintings of color. If you are in doubt whether some odd formation on a plant or leaf is a gall, split it open with your thumbnail. If there is a grub inside or evidences that there has been a grub there, you can be sure your surmise is correct.

Most galls do little harm, all are interesting once we know their history, and some have real value. The edible "oak apples," which you find attached to leaves late in summer, are galls. At one time, these oak galls were used commercially for tanning, dying and in the manufacture of ink.

Each summer, as I observe the galls forming on leaves and stems throughout my Insect Garden, the same thought comes to mind: Is the chemical injected by a gall-insect specific for one particular plant? In other words, would a gall be produced if the fly of the goldenrod laid her egg and injected her chemical not in a goldenrod stem but in the tissues of a grapevine or cinquefoil? So far, no one has been able to give me a positive answer.

JUNE • 29

A CAT HEARS A CAR A change in the wind, a shower in the night, and the morning, after days of scorching heat, is fresh and cool. As I work on the revision of a chapter, Silver curls up on the desk beside me. Nellie drives into the yard and, at the sound of the automobile, Silver leaps from the desk and runs downstairs to be fed. Cats, as well as dogs, then, can recognize the particular sound of a particular machine.

JUNE • 30

A PENTHOUSE FOR BUMBLEBEES Close to the brown swamp stream, on this last afternoon of the month, I push my way through a thick stand of cattails. At every step, my rubber boots sink into the soft mud. Overhead, the slender green leaves wave back and forth, cutting the air with their twin blades. On this late-afternoon foray into the swamp, I sniff the wind. It is scented with June flowers of the hillside. Around me, I hear the clatter of unseen marsh wrens. It is in search of their nests that I have waded into the swamp. And this search leads me to a special little adventure among the cattails.

Marsh wrens will sometimes build several nests that are never used before they construct the final one in which they raise their brood. Each is a mass of dried and yellow cattail leaves and other vegetation twined among the upright stems of a close-growing clump. Usually they are at least three or four feet above the floor of the swamp.

After finding two of the nests, newly made this spring, I come upon an old one, overlooked when I searched the swamp in midwinter to see what insects hibernated in the wren nests. Its appearance puzzles me. Like the others, it is an oval ball of interwoven leaves, six inches or so in height. But, unlike the others, it has no opening, no entrance, no doorway that I can discover. The interior seems packed solid with what appears to be fluff of old cattail heads.

Carefully I work a forefinger into the side of the nest. A little at a time, I form an opening that enables me to look inside. There are no wren eggs, no signs that birds have ever inhabited the interior. Instead, what I see are yellowish-brown objects joined together. They are smooth and slippery to my touch. They seem made of wax.

At that moment, above the swish and tap of the waving leaves, I hear a muffled droning sound issuing from the nest. At the same time, the tip of my forefinger feels a vibration. I withdraw the finger hastily. I know what I have discovered. The abandoned wren's nest has become a penthouse for bumblebees.

One of the overwintering queens apparently had come upon it in her search for a site for her colony. Usually such insect cities are established underground. But, once in a while, a queen will choose an unusual location such as the one I have discovered. Here, the queen has already formed the waxen cells in which she lays her eggs and from which come the earliest of the worker bumblebees, bees, in this case, destined to live a penthouse existence, to grow up in surroundings unknown to most of their kind.

CHAPTER SEVEN

July

JULY • 1

EARLY HARVEST New birds are everywhere—speckled robins, gray starlings, brownish redwings. Between nest and migration, the weeks are few. Birds have an early harvest. I notice that the knothole nest in the apple tree is empty; the young flickers are out in the world. I hear them calling about the orchard. How many times, I wonder, will they remember the warmth, the protection, the comforting dark of that hollow tree? Now the cavern is empty and I see pale gnats drift in and out of the opening.

JULY • 2

FISH FLY Where blue spiderwort blooms at the foot of the garden slope, I come, this morning, upon a curious, cinnamon-brown insect, lacy-winged and about as long as my forefinger. It clings motionless to the bud cluster of a spiderwort. It is the fish fly, *Chauliodes pectinicornis*. Its twin, outthrust antennae are feathered on only one side like combs. This peculiar feature gives the insect a colloquial name of "The Comb-Horned Fish Fly."

It is, I suspect, a product of the sluggish stream that flows among the cattails of the swamp. Only at night do the adult fish flies take wing. Fluttering through the darkness, the female alights on some branch or leaf or stone that overhangs the water. There she lays as many as 2000 massed reddish-brown oval eggs. The larvae that hatch from them and drop into the stream are closely related to that favorite food of the bass, the larva of the Dobson fly, variously known as the hellgrammite, the flip-flap, the alligator, the water grampus and the conniption bug. But unlike the hellgrammite, which lives under stones only in rapids and riffles and swift-running streams, the immature fish fly is often found in sluggish water. Breathing through gills, lurking under submerged logs and debris, preying on smaller creatures, it grows until it reaches a length of an inch and a half. Then it creeps up the stream bank and hides under a stone, a log or in the earth to pupate. Late in spring, the adult appears. The one I see clinging to the spiderwort is probably nearing the end of its days. For like the mayfly and the great silk moths—the Luna, the Cecropia and the Polyphemus—the adult Chauliodes lives but a short while and never eats.

THE LARGER RAINDROPS About 4:15 P.M., after this day of soaring heat, clouds pour in from the west. They close over the sky. The wind rises. The thermometer drops fourteen degrees in five minutes. Trees toss and dust fills the air. At 4:30, the first large drops pelt my Insect Garden and the storm begins. These initial drops are literally larger when a sudden summer storm commences. They descend a greater distance through the rain cloud and thus collect more moisture than drops that fall later on during the storm.

JULY · 3

SEASIDE WASPS Out on the edge of the salt meadows, in the lee of high phragmites, a great mating flight of digger wasps is in progress today. Edmund Morgan called to report the discovery. He had come upon a thousand or more of the insects, whirling and zigzagging above the dried mud of an open space close to the bay. As he watched their gyrations, he saw a strange and puzzling thing take place.

Running this way and that over the cracked mud, a baby sandpiper appeared in the opening. It had not gone far before a sudden flurry of wasps surrounded it. The dark insects alighted on it, covered its gray, downy body, massed in a ball about it. A gust of wind, sweeping over the phragmites, caught the balled insects and the whole mass, bird and wasps together, rolled over and over. Then the wasp-ball disintegrated. The insects flew about, the sandpiper chick staggered to its feet, apparently unharmed. It had gone but a short distance when the insects descended again and the whole performance was repeated. Three times the little sandpiper was incased in a ball of wasps, the ball was rolled along by the gusts, and three times the wasps flew off and the baby bird gained its feet without any sign of injury. Apparently the wasps were not stinging or attacking the sandpiper. What were they doing?

I started across the sea meadow for the spot. Even before I reached it, I caught a faint, low humming. It was a lighter hum than that of the honeybee and it suddenly rose in volume when I was within fifty feet of the mud flat. The sound was carried across the meadow by a gusty, buffeting wind from the southwest, blowing under a bright and cloudless July sky. The insects were not swarming, they were looping and zigzagging and whirling, often less than an inch above the ground. None flew in a straight line. They all seemed excited. This was a gala day. This was the great annual event upon which the future of their race depended. This was their sunlit mating dance.

The main activity centered in a space about twenty paces wide and 32 paces long. Over this area, fully 2000 dark little wasps traced in the air their endless curlicues. Beneath their whirling dance, the dry, cracked mud was peppered with what appeared to be a multitude of anthills. I walked out among the swirling wasps. They paid no attention to me. The anthills were the entrances to a host of little burrows. In them, the female wasps would place paralyzed insects to feed their offspring, born underground at the bottom of the filled-in burrows. The dancing insects were Bembex digger wasps. I counted the number of burrows in the central section of the mud flat. There were 1596. Most of them headed either east or west although a few had been excavated toward all points of the compass. Only one out of the nearly 1600 had fresh dirt at the entrance. The others all had been prepared previously.

Looking down as I walked about, I watched the whirling bodies of the wasps, spinning just above the ground wherever I turned my eyes. The movement made me dizzy. The whole earth of the flat seemed moving and unstable. As for the dancing wasps, they ignored me completely. None lit on me. None made any attempt to sting. They showed not the slightest resentment as I walked about among them. Several times, I came upon little rosettes of wasps that had alighted at one of the cracks in the mud. All faced inward like the royal court surrounding a queen bee in a hive. They seemed to be getting something from the crack. What was it, food or moisture?

More often, I encountered little balls of wasps, a dozen or so of the insects clinging together with all their bodies curving inward. Once, with the tip of my pencil, I parted such a ball and found it was clustered about a wasp that was backing out of one of the burrows. In all probability, the clusters formed about the females. At times, the gusts would catch one of the round masses of insects and send it rolling like a tennis ball across the mud flat. The movement of some of the outer wasps sometimes started the mass rolling, and, two or three times, I leaned down and, with the tip of my pencil, rolled a ball along for several inches. This kind of entomological croquet was never resented in the least by the wasps.

Over the edges of the mud flat, from time to time, a spotted sandpiper fluttered with its plaintive "Wheet! Wheet! Wheet!" It showed the concern of a parent bird with a chick hidden somewhere in the vegetation—in all probability the chick that had had the adventure with the wasps earlier in the day. It seems most likely that a female wasp had alighted on the baby sandpiper and that the other wasps had piled up around her to form the bird-and-insect ball that the wind had rolled across the flat surface of the dry mud.

JULY • 4

THE INSECTS' FOURTH OF JULY I am sitting under the oldest apple tree in my Insect Garden. It is 7:30 in the morning. Already, the sun is splashing its yellow light in the grass through every opening in the leaves. A fly, with metallic, greenish body, alights on the rim of my ink bottle. It laps up some of the black fluid and then drones away down the hillside. Here, in my garden, I am going to spend the day with the insects, with ant dairymaids and visitors to the milkweed blooms, with hunting wasps and dragonflies. Again today, as I have done in years past, I am going to set down some of the small events of the Insects' Fourth of July.

7:40 A.M. Under the curled viburnum leaves, two of the black guardian ants are still asleep beside their motionless herds of aphides. I tap the leaves and they come awake instantly. Apparently, these insects have spent the night there.

8:00 A.M. Ladybird beetles are going over the buds of the milkweeds that hang in massed green globes. But they pay no attention to the pink-white balls of the blooms that fill the air with their sweet perfume.

8:35 A.M. All over the sunward side of the viburnum clumps, damselflies cling to the pointed leaves, basking in the warmth.

9:02 A.M. I pull up a sweet flag and delight in the rich smell of the swamp that rises from the disturbed earth. A least bittern flutters away over the cattails, appearing hardly able to keep in the air. I walk back up the hillside, chewing a fragment of the sweet flag leaf. Its flavor reminds me somewhat of rhubarb.

9:50 A.M. Now the damselflies have left the viburnum leaves and are fluttering and dancing above the swamp stream. Heat is their great stimulant. Insects are like marionettes and rays of the sun are strings that move them.

10:10 A.M. A fledgling blackbird alights abruptly on a lower limb of a maple. A moment later, it begins to flutter its wings and a female redwing lands beside it and stuffs a white moth into its open mouth.

10:30 A.M. Two red admiral butterflies whirl and spin in a buffeting battle that carries them up and up higher than the apple trees.

10:42 A.M. In the top of one of the swampside maples, a dog-day harvest fly unwinds its shrilling call. It is the first cicada of this day and one of the first of the year.

11:03 A.M. Now, mingling with the intermittent call of the cicada, the creaking-toy clatter of the long-billed marsh wrens and the occasional "K-hick! K-hick! K-hick!" of a Virginia rail, there comes the faraway music of a Fourth of July band.

12:00 Noon. Grass clumps, already gone to seed, are dry and yellow at the top. The smell of hot earth and dry grass is strong all up the hillside.

12:18 P.M. The call of a mourning dove, low and mellow and seeming far away in the heat of this July noon, comes to my ears with the overtones of sadness. Yet the mourning dove is no more melancholy than the raucous jay or the exuberant robin. It, too, is a bird of courage and endurance. The gentle melancholy of its call I find in myself, not in the bird that utters it.

12:50 P.M. A small digger wasp has excavated its burrow in the hard-packed dirt beside the old barn at the top of the hillside. I notice the tread marks of an automobile tire run directly across it. Yet the burrow is solid and undamaged.

1:15 P.M. Some of the wild cherry leaves are now an almost solid mass of galls, small brown or pink ones that rise above the upper surface of the leaf on little stems like the lifted heads of miniature cobras. I pick off one of the leaves. Sitting in the shade, I count the galls. Although the leaf is less than two inches long, it carries on its upper surface 73 separate galls.

2:10 P.M. Honeybees and bumblebees, flies and ants, are feeding on the nectar of the milkweed blooms. I see one bee tug at a leg and pull it away with a bright little clothespin of pollen attached to it. Each insect that gets its feet caught in the slit traps of these flowers jerks out pollen that fertilizes other blooms, or it remains held fast by the milkweed trap. One gangling cranefly has left one of its legs behind. It had broken free from its body instead of pulling away from the trap.

2:30 P.M. The heat presses down. It lays a dullness, a growing midafternoon lethargy over the garden. Almost all birds are quiet now except for the indefatigable marsh wrens. They clatter incessantly. From one part of the swamp, where the territory of two or more wrens may be in dispute, I count twelve calls in sixty seconds, one every five seconds.

3:00 P.M. I stop beside the Katydid Jungles. This tangle of wide-leaved grass, *Panicum clandestinum*, stretches for several yards along the swamp path. This afternoon it provides a landing field for a score or more of varied insects—flies, moths, damsel flies, dragonflies, beetles, butterflies, grasshoppers, fireflies. I see them all here, resting in the heat of the sun on the green plains of the wide-leaved grass.

3:22 P.M. A hover fly hangs in the air above a little stretch of sand where wild cherry trees overarch the swamp path. Last Fourth of July and the one before that and the one before that, I saw a hover fly—standing still in the air flanked by the filmy clouds of its supporting wings—at this identical spot along the trail. It seemed to be the same fly. Yet how many generations of hover flies stretched between! The continuity of instinct, running through all the generations, had brought one of their kind each year to this same spot.

4:10 P.M. I have just looked over all the common milkweed plants, *Asclepias syriaca*, growing on the garden hillside this year. They number more than half a hundred. Yet on only one milkweed do I find any aphides—a pale, greenish cluster on the underside of the tip of a single leaf.

4:45 P.M. A female monarch butterfly crosses the garden, flutters above the milkweeds, alights, takes off again. It comes back to the milkweeds several times, apparently laying eggs.

5:20 P.M. I stand a long time watching a carpenter ant where woodbine ascends the trunk of one of the maple trees. The black, foraging insect follows the stem upward from one alternately branching leaf to the next, running on a zigzag trail over each leaf to the very tip and then back to the stem again. It occurs to me that a valid simile is: As thorough as an ant.

5:52 P.M. Wild onions have gone to seed. The pointed parchment-like envelope that encloses them is perched at the top of the green stem like some little Moorish minaret.

6:00 P.M. Thirty-eight young starlings have congregated in the dead top of one of the swampside maples. They are flocking already. A sign of autumn so early in July!

7:10 P.M. All day today I have seen unusual numbers of the black firefly, *Lucidota atra*. They have been fluttering among the tangles of vegetation, their slow and feeble flight resembling that of caddis flies. Now, as evening comes on and I prepare to leave, they are more in evidence than ever. Never before can I remember them so numerous. Each Fourth of July that I spend on this same hillside I see, *in general*, the same things, the blackbirds, the damsel flies, the carpenter ants and the sweet flags. But, *in particular*, no two years are ever the same.

JULY • 5

DEATH AND DESTRUCTION In nature, there is less death and destruction than transmutation.

JULY • 6

THE ANT'S DAIRY This morning, when I looked again at the milkweed leaf holding the aphides, I found the colony had expanded and several red ants were collecting honeydew. On the other leaves below, other red ants coursed about. All the ants on the milkweed were the same kind. Here was a chance to test something I had wondered about. Are the aphides on a bush or plant visited by ants from various nests or are they the property of one colony? Does each mass of plant lice, guarded by ants, represent the dairy herd of those particular ants? I soon had my answer. Picking a black carpenter ant from another aphid-infested plant, I dropped it on the milkweed. If different colonies collected the honeydew, the new ant would not be molested; if the aphides represented the red ant's dairy, it would be attacked. It was. Immediately, the red ants went on the offensive and the black ant ran wildly about as though seeking escape. It fled downward to another leaf. There it came face to face with a red ant which fell upon it instantly. A second red ant joined the struggle. All three rolled about until they tumbled to the leaf below. Here the black ant broke free. It ran down the stem, through the gantlet of other attackers, and finally escaped in the grass. This particular milkweed was the dairy, owned and defended by the red ants. Apparently, all such sources of food, whether they are fallen fruit or herds of honeydew-producing aphides, are the property, jealously guarded, of one particular colony of ants.

JULY • 7

THE HEAT OF THE DAY To Milburn Pond a little before noon. No breeze stirs. The cloudless sky seems a vast blue flame pouring its heat on the earth. I walk down the pond edge, under the glare of the sun, noting the changes since my last visit. The mallard ducklings, that I saw so short a time ago darting about like water bugs, are now half grown. The pebble patches where the bream eggs hatched are silting over again. The yellow flowers of the water iris have come and gone and the cinnamon ferns, whose fiddleheads I saw unrolling, now stand in dense clumps and ranks in the muddy lowland spots. There is never a pause, on summer days, in the changing kaleidoscope of nature. I turn out of the heat into a path under the trees. The cool woodland air is as welcome to my lungs as the filtered green woodland light is to my eyes. This is my thought at the moment. It is, to use the words Andrew Marvell, the English poet, wrote more than two and a half centuries ago, "a green thought in a green shade."

JULY • 8

BARN SWALLOWS At nine o'clock this morning, I found the nest in the wagon shed empty. Yesterday, the four baby barn swallows filled it to overflowing. Early this morning they fared forth on their first flight and I see them now, ranged on the dead twigs of an apple bough being fed by the parent birds.

It was on the 21st of May that I noticed the first wet mud cemented to the side of the roof beam. By the 26th, the cup nest was done. In the days that followed, I saw white chicken feathers carried to line the cup. Two seemed a bit too long and stuck up at the back like the feathers in an Indian's hair. During the early days of June, there was always a swallow on the nest. In incubating the eggs, the male and the female take turns. I would see the relief bird arrive with a twitter and slip in one side of the nest while the brooding bird left by the other side. The eggs within were exposed to the air for only the briefest interval. This changeover occurred about every fifteen minutes and the relieved bird headed down the slope to the swamp where it fed on flying insects during its time of freedom. In the air, the work of the swallow seems fun, its food-getting an art, a kind of aerial ballet.

Few birds are as peaceable and sociable as the barn swallows. At times I would find other swallows in the shed, apparently visiting. When the parent birds perched on a twig or on a rope at the back of the wagon shed, they would move closer together until they were almost touching, twittering in a wide variety of liquid calls. Occasionally, when resting alone, one of the birds would stretch out one wing and yawn like a sleeper awakening. For all their peaceful ways, however, the barn swallows are brave in time of need. Without talons or beaks, they dive on cats and dogs and even humans when the young are ready to leave the nest. I saw one bird set out in pursuit of a swallowtail butterfly as it fluttered past the open door of the wagon shed.

During the last days, the four fledglings stuck out over the edges of the nest. It became a cup running over. When a parent bird entered the shed with food, its twittering caused the quartet to shrug and push until they all faced outward with open bills. Now the open bills are gone, the nest is empty, the twittering within the shed has ceased. But I hear the sound outside and turn to watch the new swallows on their first day in the air.

JULY • 9

COTTON TAIL I was walking slowly across the hillside in the cool of the evening today when a baby cottontail bolted out of a grass clump at my feet. It was probably one of those whose little ears I had seen in the fur-lined nest on this same slope above the swamp. The young rabbit shot for the path, made such a sharp turn its feet flew out from under it and it pawed the air frantically. It got its footing, bolted away in another direction and, in its panic, tried a second right-angle turn. Again it upset and slid on its side with flying feet. Three times it shot off. Three times it turned sharply. Three times it upset before it disappeared in the course of its helter-skelter, zigzag, panic-stricken flight.

JULY • 10

JULY DROUGHT Rainless days of heavy heat have scorched the grass of the hillside to a toasted yellow. The leaves of the milkweed and the butterfly bushes droop in the dry air under the blaze of the sun. Beside the swamp stream, an edging of brown is creeping along the sword leaves of the sweet flags. Summer already seems slipping away, seems beginning its long decline. Each year, after the Fourth of July, how fast the world spins on its axis!

JULY • 11

CHICORY BLUE On this day, I drive to Concord, Massachusetts, for the annual Thoreau Society meetings tomorrow. Along the way, I see that most beautiful of blues, the tint of the wild chicory in bloom. It seems to me the most perfect blue on earth, the most perfect blue under the sky.

THE CONCORD ANT Across the street from the Colonial Inn this evening, I see a Concord ant engaged in Herculean labors. I first notice the white wing of a moth traveling across the sidewalk, seemingly supported by magic. Then I observe a small, dark ant. Although it is less than one-eighth the length of the moth wing, it is gripping the base of the wing in its jaws, holding its burden straight out in front of it, and thus running at top speed across the concrete of the sidewalk. How fabulously strong must be the neck of such an insect.

JULY · 12

A PAPER NAPKIN Before breakfast at Helen's Restaurant, on this anniversary of Henry Thoreau's birth, Raymond Adams, Ira Hoover and I walk to the Sudbury River and back under the great elms of Concord. Once, we saw what appeared to be a white dove fluttering down into a yard. We passed beyond some bushes and saw that it was not a dove but a white tissue paper napkin drifting down from a height of fifteen or twenty feet. It had hardly touched the ground when a robin alighted beside it, pecked at it, grasped it in its bill and flew laboriously upward to the limb of a maple tree. Here it rested a moment, then ascended another ten feet to another limb. In the deep shade, the dark robin was almost invisible and the shining white napkin seemed rising by itself. In all probability, the paper was on its way to provide another instance of "something white in a nest."

JULY · 13

THE RIVERS OF CONCORD I stay over this extra day after the meetings are ended, to follow once more, as in years past, the beautiful rivers of Concord. For eight hours, I paddle a canoe along the winding waterways of the Sudbury, the Concord, the Assabet. I ride through duckweed rafts, under the Leaning Hemlocks, past Bittern Cliff, into Fair Haven Bay, beside the Great Meadows. The reach of the river exceeds the reach of the road. The rivers carry me to a wilder Concord than I could attain by any highway.

Wherever I go, along the stream edge the calm, flat water reflects the maples and the elms. Ripples produced by my moving canoe are the winds of the reflected trees. They set their boughs to waving. In this dark mirror of the river, dragonflies appear inverted and they increase or diminish in size as they rise or descend above the surface. Everywhere on the three rivers today, damsel flies are laying their eggs. Each floating piece of waterweed or leaf is freighted with the insects. Near the Leaning Hemlocks, on the Assabet, I come upon a foot-long strand of waterweed caught on a snag. It supports the close-packed bodies of thirty clinging damsel flies.

Mile after mile, the shallow outer edge of every bend of the winding Concord is alight with the purple-blue flames of the pickerel-weed. Once I push my canoe into a swampy bay where the *Pontederia* rise all around me and where snuff-colored dragonflies are dipping to lay their eggs. How kitten-soft are the unopened buds of the pickerel-weed! Cattails here are interlaced with the bur-reeds, their green seed balls resembling round and stubby gherkins. Bumblebees visit the white globes of the first flowers on the buttonball bushes and the red-purple of the swamp milkweed, here and there along the riverbank, shines out amid the lush vegetation. I hear goldfinches flying over the Great Meadows, killdeers off on the Bedford Levels, veeries singing among the lowland maples. There is the rattle of a kingfisher and the trump of a bullfrog. I see holes in the river bank and they excite me as though I were a weasel or a ferret.

From time to time, I let the canoe drift into shallows where white pond lilies have opened waxen petals, revealing centers that seem plated with gold. I lift and smell one of the dripping flowers. Instantly, my nose is filled with a cloud of minute flies or gnats that have been feeding unseen on the nectar. What am I reminded of by the perfume of this river flower? It seems to me there is in it the faint suggestion of licorice. On scores of the widespread lily leaves there are clinging the nymphal skins of emerged dragonflies. They poise there like tiger beetles ready to take off. I see one dragonfly skin anchored to a *Pontederia* leaf with the egg case of a spider clinging to it. Nothing stands alone in nature. The supported becomes the supporter. Each brick rests upon another brick.

At noon, as I am drifting among pond lilies eating my sandwiches, a painted turtle climbs onto a log, dripping and speckled with green duckweed. It catches sight of me and tumbles off into the river. A moment later, the tip of its head appears, then winks under the water again, the surface snapping shut over it like a blinking eyelid.

Paddling slowly back against the current, I round a bend in the river and come upon a wood duck and seven ducklings disappearing among the pickerelweed. At the next bend, I look back. Duck and ducklings are out on the river again. Ninety years before this July day, Henry Thoreau followed the Concord River to the Great Meadows. He noted how the dry heads of the Canary grass rose above their green leaves, how the fruit of the bur-reed was "pickle-green," how the lily pads were much eaten by insects, how the leaves of the *Pontederia* were slit into ribbons—he thought, perhaps, by some bird. These things are the same today. I see the dry Canary grass, the green bur-reed fruit, the pads nibbled by insects. I, too, see pickerelweed leaves slit and am at a loss to explain them. After nine decades of vast and worldwide change, how little changed are these banks of the Concord!

JULY • 14

MOLTING HENS Driving home today, I followed—for a mile or so—a truck piled high with chicken crates. Each crate was crammed with white hens. Following the truck was something like driving through a blizzard for the air seemed filled with white, flying feathers from the molting hens. I wondered if the change and excitement increased their molting as excitement will increase the shedding of hair by a cat.

THE SNAIL'S PACE It takes days of practice to learn the art of sauntering. Commonly we stride through the out-of-doors too swiftly to see more than the most obvious and prominent things. For observing nature, the best pace is a snail's pace.

JULY • 15

CUCUMBER SPRING Back on my garden hillside at dawn on this mid-July day. I notice how a wild cucumber vine has climbed upward from branch to branch in a low cherry tree. Tendrils, spiraled like a coiled bed spring, secure it to each limb. I pull on one of the tendrils and note with interest its resiliency. Here is nature's patent model for the shock-absorbing spring. On windy days, how often the living springs of the tendrils must play a part in saving the vine from being carried away by gusts.

JULY • 16

RUBY-THROAT It is midmorning. The heat is mounting but some of the freshness of the night is still in the air. While I was away, rain relieved the drought of the hillside. I sit in the grass, leaning back against the bark of an apple tree, thinking of nothing in particular, letting my gaze wander about over the hillside. As I sit there a sudden airy swish of a sound reaches my ears. A ruby-throated hummingbird has arrived among the buddleia blooms of the garden. All along the Atlantic coast and inland beyond the Mississippi—two-thirds of the way across the continent—there is only one species of hummingbird, the familiar and remarkable ruby-throat.

It can take off in seven-one-hundredths of a second. It can hover like a helicopter. It can fly backward. It can pass an automobile traveling down a straight road at fifty miles an hour. Its wings, carrying its thumb-sized body through the air, move as fast as seventy-five beats a second, 4500 a minute.

In recent years, ultra-high-speed motion pictures have revealed for the first time exact information about the movement of the wings of the ruby-throat. To our eyes, they form little clouds that shimmer and blur on either side of its body. But the Edgerton camera has slowed down these vibrating wings on movie film so that scientists have been able to count their movements. They have found that when a hummingbird is hovering, its wings beat 55 times a second; when it is moving backward, 61 times a second; when it is flying at top-speed straightaway, 75 times a second. If, as has been frequently said, ruby-throats fly over 500 miles of open water in crossing the Gulf of Mexico during migration, their tiny wings must beat more than 2,000,000 times without a pause!

Now, scattered across more than half the United States, ruby-throats are using their wings to fly from flower to flower, as the hummingbird in my garden has done. They are using them also in transporting plant fibers, spider's silk and bits of lichen for the building of the thimble-like nest and in carrying food to the two naked, pea-sized baby birds that hatch from the tiny snow-white eggs. They also are using their vibrating wings in valiant, rushing attacks on interlopers. A hummingbird will dart at a bumblebee, at a butterfly, even at a kingbird. Once one was seen flying at top speed in pursuit of a chimney swift.

Early settlers in the New World debated for a long time over whether the ruby-throat was really a bird or a "West Indian bee." The first naturalists who sent back museum curiosities to Europe tried to press hummingbirds in the manner of botanical specimens. Vibrating with life, scintillating with color, the ruby-throat, wherever it is seen, darts through the air with the bright perfection of a feathered gem.

JULY · 17

OLD CROW AND OLD OWL Why do people so frequently say: "There's an old crow on that fence" or "An old owl lives in that tree"? Why "old"? It is probably by way of being a compliment. Is it not because we think of such birds as being old and wise?

JULY · 18

SCUM After the heat of the day, Nellie and I walk across the sea meadow to the bay and eat our supper of sandwiches, once more, near the storm-stranded windrows of seaweed. Now, the clapper rails call less frequently. The seaside wasps have long since concluded the whirl of their mating dance. Their burrows are filled in; the mud flat is deserted.

As we walk back over the salty cordgrass and the green samphire, we stop where a small depression in the sea meadow forms a tiny scum-bordered pool. In the lessening light, on that wide stretch of darkened land, the puddle seems to catch and concentrate all the glow of the evening sky. A band of brilliant green edges the pool and tiny flies land and take off on the surface of the glowing water. As I gaze down on this algae-bordered puddle, a kind of despair envelops me. What are the algae of the green border? I don't know. What are the little flies landing and taking off? I don't know. What are the small plants thrusting up through the water? I don't know. The vistas of my ignorance seem boundless. How much that I see I do not recognize; how much that I observe I do not understand! In this despairing and humbled mood, I traverse the meadows and return home. In the study of nature, we never exhaust the possibilities of an area; the area exhausts the possibilities in us.

JULY · 19

THE SEEING EYE Thinking of that sea meadow pool and the mood it evoked last night, I have been remembering a time when my Insect Garden was invaded by experts. Between thirty and forty of them arrived for the annual outing and field trip of the New York Entomological Society. They had brought their naturalist friends and some came from as far away as Philadelphia. Each in his own way saw a different facet of the garden activity.

Beetle experts began peering under bark and lifting up old logs and making a harvest among the goldenrod of the hillside. Butterfly enthusiasts worked the open fields and moth-men hunted among the wild cherry leaves for unusual larvae. A mosquito collector immediately investigated all the knotholes in the old apple trees. The life of the swamp stream absorbed one entomologist; a termite nest under a decaying board interested a second; leaf-hoppers among the sunflowers, a third. There was an authority on mayflies, another on wasps, another on stink-bugs. There was even a snail collector in the group.

The thing that fascinated me all day long, as I watched these friends of mine in action, was the functioning of the seeing eye. Side by side, two of the specialists would look at the same bush. One would see leaf-hoppers, the other slug caterpillars. They were concentrating on their special fields, seeing mainly what they were interested in seeing. Although I have known this area intimately for a decade and a half, have walked back and forth and up and down its paths far more than the equivalent of a pedestrian tour from coast to coast, I had missed many of the things these men and women saw on their short visit. Their specialized knowledge gave them seeing eyes I did not possess. In truth, no matter how much we know, there will always be someone who will see what we overlook, someone who will understand what we gaze upon without comprehension.

The snail-man left with an amazing collection of tiny shells from along the edge of the swamp—shells I did not know existed there. And the student of wasps, with the aid of a wad of ether-soaked cotton batting, unearthed a whole colony of yellowjackets from the open field north of the apple trees. He left with the colony in a shoebox which he carried gingerly under one arm and around which he wound a prodigious amount of string. Later he reported that his first impression was correct. These particular yellowjackets had never been recorded as nesting in New York State before. And all that summer I had been passing the entrance of this colony, watching the insects come and go. I saw they were yellowjackets and that was all. It takes the expert, and the seeing eye of the expert, to recognize the unusual.

JULY · 20

INDIAN SHELLS In the field south of the garden, there is an irregular ring, seven or eight paces in diameter, where vegetation is less luxurious and where evidences of drought first appear. The cause of this phenomenon goes back to Indian days. Just below the surface of the ground, the ring is formed of a solid mass of clam shells. Apparently, the Indians sat at their campfire and tossed the shells back around them until the ring was formed. It has been part of this field as long as the oldest resident of the community can remember.

JULY • 21

THE MUGGY DAY The night had no coolness in it and the dawn is heavy and humid. Everything is moist and misty. The air seems turning from gas to liquid. Walking about my garden hillside, I come upon a dozen or more tan-colored garden slugs moving across the milkweed leaves. Behind each, a varnished trail glistens in the misty light. Damsel flies, their wings spangled with droplets of moisture, cling to the top of the arcs of the down-drooping flags. After nights of heavy rain, I often find the damsel flies clinging beneath the leaves. They have, so to speak, drawn a green coverlet over them while they slept. They have gone under a leaf-tent for a rainy night.

I look back at the wild cherry tangle. How swiftly the fruit is forming! Where there was only the white foam of flowers but a short time ago, there is now the massed green clusters of the developing cherries. Out across the marsh, silvered by the mist, new, light-brown cattail heads are rising in straight lines amid the tumbled sea of slender leaves.

The sun rises. The muggy heat increases. New robins, with speckled breasts, fly a little awkwardly among the dead upper branches of the apple trees. Where the willow overhangs the swamp stream, I see a redwing, another young bird not long out of the nest, hopping from limb to limb. It has the awkwardness of youth; it is not quite sure of its balance; it shifts its feet with difficulty; it is learning by practice. These days are avian school days all across the land.

Moist and comfortable on this humid morning, almost like froghoppers within their castles of foam, the plant lice are placidly drinking sap and giving birth to a seemingly endless succession of living young. Uncounted generations follow each other through the summer days. I stand for a long time contemplating the swiftness of childhood and maturity and old age among the peaceful aphides. I look about me. Everywhere there are the new generations of another year—new apples, new robins, new flowers. Each plays its part in the unending repetition of the seasons. At length, I rouse myself from this contemplation of life cycles so much shorter than my own. I turn toward home with the feeling of having lived for a time the life of some ancient Methuselah watching the flow of generations passing by.

JULY · 22

BATTLE BY BOUNCE Two bluejays in the backyard are quarreling over food. As each rushes toward the other, it bounces high in the air. This is a battle by bounce. The birds seem buoyant, as though partly filled with helium.

JULY · 23

REVERSIBLE INSTINCTS It is a strange story that scientists have revealed about the gold-tailed moth of Europe. In the spring, when the earliest green is on the twig tips, the larvae of this moth appear from hibernation near the ground. Light attracts them powerfully. Instinct leads them to get as near the source of the light as possible. They crawl upward until they come to the very ends of the upper twigs. And it is precisely there that the tenderest buds provide them with the food they need. Their instinctive attraction to the light draws them irresistibly to the only available source of food. They follow a direct path instead of wasting time in an aimless search.

So powerful is this attraction that a caterpillar put in a glass tube, with one end facing a light, invariably crawls at once toward the lighted end. In fact, when choice buds are placed in the darkened end, the larva is unable to go to them although it may be dying of hunger. It remains charmed by the light and dies of starvation within a few inches of plenty.

But—and this is an even more curious feature of its story—once the larva of the gold-tailed moth has devoured the twig tip buds, it is no longer attracted by the light. Instinct reverses itself. The caterpillar is now repelled by the light. It gradually works its way down the twigs. If this were not so, it would be held among the bare upper twigs by the attraction of the light and would be unable to descend and feed on the lower leaves as they unfolded. Devouring the first few buds seems to release it from its spell. After that, it can crawl in any direction.

Birds, I suppose, exhibit a kind of reversing of instinct in their nesting and migration. At their breeding area, the roving instinct is replaced by an almost sedentary one; their instinct to travel is supplanted by an instinct to stay in one place. This, in turn, gives way to the roving instinct again.

Another instance of an instinct reversing itself is found in the life of the cluster flies, those insects that sometimes gather in close-packed groups on the inside of windowpanes in country schools in the fall. During summer, these insects seek out sunny spots. But as soon as the temperature falls much below 60 degrees F., they are repelled by light. They creep through cracks into the darkened interior of buildings. Then, as soon as the interior temperature rises above sixty degrees, they are once more attracted to the light and fly to the windows to gather in clusters there, held irresistibly by the greater brightness of the light outside.

JULY · 24

A SUMMER THOUGHT There are certain thoughts that keep recurring in our minds. Each year, during sweltering summer days, the same reflection occurs to me. I remember, with a sense of wonder, how difficult it will be to recall my sensations in the heat of July when—only six months hence—I am amid the cold and snow of January.

TREE CRICKET At 6:44 P.M. as I sat at my desk entering the above words, I heard, just outside my window, one of nature's most beautiful sounds, a voice of other years so far unheard this summer. It was the long, sweet, mellow trill of the year's first snowy tree cricket.

JULY • 25

THE ROBIN AND THE ANT Under a maple tree in our backyard, a robin is hopping about, its tail twisted at a peculiar angle, its wings hanging down, its head turned far to one side. Suddenly the bird pecks at the ground, lifts one wing and rubs its bill vigorously on the plumage underneath. It hops a foot or more, pauses, repeats the performance. This goes on for five minutes. Over and over again, the robin picks something from the ground and rubs it on various parts of its body. Then it rises quickly and disappears over the boundary fence.

What has it been doing? It has been engaged in that peculiar rite practiced by many birds and known as "anting." The ground under the maple tree is riddled with the nests of ants. Each time the robin pecked at the ground, it was picking up an ant. We have seen other birds—starlings, red-winged blackbirds, bluejays and grackles—come to this same spot for the same purpose. Half a hundred different species have been observed engaging in this rite.

The assumption of scientists is that the birds are using the ants to rid themselves of body lice. But just what happens is still in debate so far as I know. Do the birds put the live ants in their plumage or do they rub on the insects to get formic acid as a dressing or as a repellent for the lice?

Even though we have watched the birds in our backyard closely through field glasses all during their anting we have been unable to decide which method is used. It is impossible to tell whether the bird, in its vigorous rubbing, is working its bill down into the fine plumage to place the ant in contact with its body or whether it is rubbing on formic acid. If the formic acid is used as a repellent or insecticide, it is curious that, so far as I know, no bird resorts to the nettle as a source of this fluid.

The sting of the nettle is due largely to formic acid. Small sacs within the plant are under tension so that when the tips of the nettle needles are broken off the formic acid is forced out. This causes the smarting that is familiar to anyone who has mistakenly grasped the head of one of these plants. Here is a supply of formic acid ready for the taking. Yet I know of no observation of a bird applying the fluid from a nettle to its feathers.

JULY • 26

SIXTY-THREE ANTS Today it is a starling, instead of a robin, that is anting in the backyard under the maple tree. But its movements and positions are as odd and ludicrous as those of yesterday's redbreast. Its stubby tail points off at a sharp angle to one side. It twists and tilts. It assumes awkward and unnatural positions. It seems deformed and distorted and falling apart. An anting bird, by its curious stance and behavior, can be recognized at a glance. Nellie watches this starling from the moment it lands until it flies away. During that time, it pecks at the ground and rubs its bill on its feathers sixty-three times. Presumably it uses a different ant each time.

JULY • 27

A SQUIRREL IS AFRAID OF SWEET CORN This afternoon I tossed an unused ear of sweet corn into the backyard for the birds. A squirrel found it. It was obviously the first corn on the cob it had ever seen. The smell attracted it; the unfamiliar object aroused its caution. It came in from all sides. It approached and darted back. It flattened itself on the ground, inching forward, nose outstretched. It was tense like a taut spring, ready to explode into action at the first hostile movement of the ear of corn. Once it almost touched the ear with its nose only to leap sidewise in a sudden panic and dash halfway up a maple tree. The ear fascinated it and drew it back time after time. For half an hour, it kept returning without touching the corn. Then, in a great triumph over indecision and alarm, it grabbed the ear and rushed away.

JULY • 28

THE UNIVERSAL SOLVENT The universal solvent is memory. It dissolves the past. It eliminates time. A moment ago, as I wandered among the sunflowers of my Insect Garden, some fragment of sound, some passing emotion, recalled a time in the Indiana dunes fully forty years ago. For the moment, forty years ago is closer than yesterday. That remembered sunset over the Lone Oak hills when I was a child is more real, more immediate, than all the sunsets of all the years between.

JULY • 29

AN INSECT PUZZLE After a day in the city, to the garden at sunset. The air is sweet and cool. Sunshine and shadow streak the trunks of the old apple trees. Little flies whirl and dance in the low rays of the sinking sun—flying sparks from the fire of life. The voice of a cicada soars and then sinks away to a metallic snore in one of the maples. Craneflies bump and blunder amid the grass tangles, and an English sparrow hops about among the stems of the roses, feeding on aphides.

Where it is touched by the rays of the setting sun, I watch a black carpenter ant on a milkweed plant. It runs about a leaf, this way and that, starting and pausing, running in circles and curlicues, all the while vibrating its body in a vertical plane. It seems actuated by a lopsided cog. Its progress is erratic. It runs with a kind of limp or catch. However, irrespective of the speed of its progress, its body continues to move rapidly up and down. The shadow of its abdomen is cast on the leaf by the setting sun. It moves up and down, now like a clicking telegraph key, now swiftly like drumsticks beating a rapid tattoo. It never ceases its vertical movements so long as the insect is running or walking. Several other carpenter ants are on the milkweed but this one can be recognized instantly. Wherever it goes, on one leaf after the other and on the stem as well, its peculiar gait distinguishes it.

At first I think it may be laying down a formic acid trail. But its shadow shows that the tip of the abdomen never touches the plant. It is moving up and down in the air alone. Other ants, passing by, pay no attention to it. Is it a freak, an ant with some kind of nervous affliction, some kind of tic or mannerism? Will I ever know?

JULY · 30

A DRAGONFLY MOVES BACKWARD I spend an hour today on the edge of the swamp stream watching dragonflies in a breeze. If I could see one of them move backward *against* the breeze, it would supply the final answer to the question: Can a dragonfly fly backward?

The heat of the midday sun pours down on the swamp. The breeze, slight but steady, is filled with the heavy smell of vegetable decay. All around me, dragonflies course up and down the dark lowland stream. They hang in a glitter of wings. They dip and zoom. Under the July sun, life for the *Odonata* is at a peak. In an ever-reforming procession, the insects dart past me with the rattle and flutter of their parchment wings.

One alights on the drooping leaf of a sweet flag hardly four feet away. It is brownish with spotted wings. It is one of the skimmers, the Tenspot, *Libellula pulchella*. It clings there in the sun, its wings glinting as it rocks slightly in the breeze. I change my position to get a better view. The bulging, compound eyes of the dragonfly instantly catch the movement. It shoots upward from the leaf. Then, watching from the side, I see it hover for an instant before it suddenly moves several inches to the rear. This time, there is no question of a breeze moving it. For its backward flight carries it directly against the flow of air. Twice more I see this same *Libellula* make quick backward movements, once for only an inch, the other time for a longer distance. But, in each case, the movement is not with the wind but directly against it. Thus, after months of intermittent watching, I have the final answer. Dragonflies fly backward.

JULY · 31

THE FALLING FEATHER About two o'clock on this hot last afternoon of July, as I am writing in the shade of one of the old apple trees, a molting bluejay flies past down the garden path. It turns sharply, with tail widespread, and alights close to the sunflowers. I notice the tail has lost a feather. Spread wide, it resembles a smile with a missing tooth.

After a minute or two, the jay takes wing again. It climbs steeply. And, as it climbs, a downy breast feather falls away to ride on the thermal updrafts. It drifts through the air like a little blue cockleshell boat, turned up at either end. All across the open space, it is supported by ascending currents of heated air. It rises and falls on invisible waves. Eventually its drifting course carries it above a low clump of young wild cherry trees. Instantly, it dips and drops toward the leaves. Cooler air is here. The updrafts have ended. Each bush and tree is, in a manner of speaking, a green fountain spraying moisture from all its leaves. In consequence, the air near the leaves is cooled. This I notice on summer days of heat. Whenever my bare arms brush against wild cherry leaves along the path, I notice the sudden coolness of their touch.

CHAPTER EIGHT

August

AUGUST • 1

SIGNS OF CHANGE Two butterflies spin round and round above the buddleia bush. I watch them disappear over the wild cherry trees, still whirling on an invisible axis like some heavenly body moving through space. The late-afternoon air is hot and still. I sit in the grass, leaning back against the trunk of the oldest apple tree—the one I call the Lincoln Tree because it was planted when Abraham Lincoln was president. Idly, I follow the progress of a carpenter ant up the trunk and out the underside of the lowest branch. I notice how the bark, where branch and trunk meet, is wrinkled like the inside of the elbow of an unpressed coat. Then I become aware that far above the tree, in the pale blue sky, there is an endless flashing of white. Little glintings appear and disappear continually. They are the white breasts of tree swallows turning and catching the sun. Already swallows are moving down from the north. I saw them ranged side by side on telephone wires this morning. In the circle of the seasons, there is no pause. Already summer slides toward autumn. On this hot afternoon, at the very summit of the season, signs of change are in the air.

AUGUST • 2

SQUIRREL TAILS In recent days, I have noticed how the banner tails of the gray squirrels have suddenly become threadbare. Their bushy tails are bushy no longer. Much of the hair has been shed to make way for new hair. Because of this, these midsummer days may be a time of danger for gray squirrels. Perhaps they have to be more careful now in launching themselves from tree to tree. For, normally, they depend upon their bushy tails for balance. Bob Hines, the wildlife illustrator, once told me of a gray squirrel he used to watch in the White House grounds in Washington. Through some accident, it had lost much of its banner tail. It never leaped across gaps between trees, as did its companions. Even when pursued by another squirrel, it managed to avoid long jumps in which the absence of a balancing tail could prove disastrous.

AUGUST • 3

THE SNAPPING LINK In the hazy noontime heat of this third day of August, I wander along the swamp trail. Gone are the drifts of blackberry blooms. In their place, I see the ripening berries, so shining they seem made of enameled pottery. A baby snapping turtle, no larger than a silver dollar, is floating on a tangle of waterweeds in the swamp stream. Rushing past me, two dragonflies, with frosted-blue bodies, swerve among the sweet flags and I hear the dry rattle of their wings as they thread their blundering way under the drooping leaves. I turn away. And, as I turn, I catch sight of little ripples running over the dark mirror of the stream. A green darner dragonfly has become entangled in waterweeds near the bank, there it lies, trapped on the surface, spasmodically vibrating its wings. I fish it out on the end of a stick and lay it on an arum leaf in the sun. In a few moments, it should be darting through the air again. But when I look at it, a few minutes later, it is motionless, dead, perhaps exhausted by its struggles.

Before I came upon its little tragedy, a sentence from Adrien Le Corbeau's *The Forest Giant* was running through my mind: "Time then is not to be divided into centuries and hours and minutes, but just as each atom figures a world in itself, so each moment is an eternity, one link in an endless chain which binds the remotest past to the remotest future." So too, I had been thinking, the chain of life connects past and future. The days of the green dragonfly had formed one little link in that vast chain. Its death, in this time of hazy heat, was but one of unnumbered, unnoticed little tragedies of the summer days. A tiny link had snapped. But, even as I stood there, two other green darners, flying in tandem, alighted at the base of a sweet flag. The lower insect, the female, dipped the tip of her abdomen into the stream again and again, depositing her hundreds and thousands of eggs in the water. The links snap but the chain goes on.

AUGUST • 4

FRIEND OF THE KILLDEER A pair of killdeer nested this spring in a coal yard a mile or so from here. One of the workmen took a special interest in the welfare of the birds. He protected the nest. He kept everyone away as the time of hatching drew near. Once the baby birds appeared, he devoted himself to keeping the little family together. He was unaware that young killdeer can run almost the moment they step from the egg and that they normally scatter and hide for safety. He thought the little birds were getting lost and tried to put them all back in the nest again. Around and around the coal yard he ran in pursuit of the fleet-footed chicks. They hid behind pieces of coal. They scuttled this way, darted that. At the end of the day, their benefactor was worn out and thoroughly convinced that the family was irretrievably scattered and lost.

AUGUST · 5

NATURE'S CALM After the heat and fumes and tension of a city day, I came to the green peace of my garden hillside. Nature has her storms and hurricanes and lightning. But, as year follows year, her prevailing mood is one of steadying calm. Hours spent in the open leave us less tense, more at ease, as though by association we had obtained some of nature's calm assurance.

The green of the grass and trees is restful to our eyes. The sound of the wind and water is calming to our nerves. We lift our eyes to the mountains or we gaze out over the vast blue plains of the sea or we lie on the sand of the beach in the darkness and stare up at the whole heavens sown with stars and we know something of natures underlying mood.

The rustle of leaves, the ripple of brooks, the lap of waves, the whisper of grasses when the breeze runs across them—all these are relaxing sounds, calming sounds. Our ears are hungry for them. They help remove a little of the tautness from modern living. Amid such things, we feel ourselves losing tension, relaxing like the drought-dried plant in a summer shower. This is the great gift of nature's calm.

AUGUST · 6

THE ANTING BLUEJAY At 8:15, this morning, a bluejay flew into the cedar tree next to the kitchen window. It began to hop about, looking intently at the leaves of a young maple that had sprouted beside the cedar and now rose to a height of five feet or so. The bird pecked at one of the leaves and then lifting its left wing rubbed its bill vigorously on its plumage. It pecked again and repeated the performance under its right wing. On the limb of the tree, several feet above the ground, the bird was anting.

Eight times it picked ants from the maple leaves and rubbed them on its feathers. Then it flew away. I examined the maple after it had gone. Under many of the leaves green-and-brown aphides clustered. They were attended by carpenter ants. On the 106 leaves of the little sapling, I found 96 guardian ants. Directly beneath the spot where the bird had perched, I noticed several injured ants. They were curled up with their legs moving feebly. As I watched, one died, another recovered completely and ran off toward the base of the maple sapling. The others appeared to be reviving. So the anting bird does not leave the insects in its plumage, as some had supposed. Neither does it crush up the ants and rub them on its feathers, as others had suggested. The jay had apparently held the ants in its bill and by squeezing, or merely by the act of holding them, caused them to give off their formic acid. Then it had dropped them to the ground.

A POSTSCRIPT ON ANTING Not always does the anting bird use ants. Instances have been recorded in which several grackles rubbed orange peels on their feathers. At other times, these same birds anointed themselves with juice obtained from the husks of walnuts. Starlings have been seen rubbing on vinegar and lemon juice and beer. One tame magpie was in the habit of dipping ants in the tobacco ashes in a pipe before rubbing them into its plumage. There are several records of birds that used cigar butts in their anting. In her classic studies of the song sparrow, Margaret Morse Nice saw one male bird rub sumac berries under its wings and on its tail feathers. Choke cherries, apple peels, the rind of limes have been employed by other birds. An Australian naturalist named Givens is reported in Frank W. Lane's *Animal Wonderland* as seeing a flock of a dozen birds or so using smoke in their anting. Daily, these birds, red-browed finches, visited a smoldering log where acrid smoke curled up from small cracks in the bark. The finches would thrust their bills into the smoke and then rub them vigorously on their feathers. Not long ago, a correspondent in York, Pennsylvania, wrote me about blackbirds and grackles in her backyard. To discourage skunks from rooting up the lawn, she was in the habit of scattering moth balls in the yard. Rain reduced the size of the balls and when they were small enough for the birds to grasp them in their bills, the redwings and the grackles would pick them up and, lifting their wings, rub them on their feathers.

AUGUST • 7

DO INSECTS PLAY? A robber fly has alighted close beside me on a wild cherry leaf as I lean against the Lincoln Tree. It recalls a question about insects. Do they ever engage in play?

Stanley W. Bromley, who has spent many years in the study of these predaceous flies, once told me of seeing an occurrence that had all the earmarks of an insect at play. Eight or ten miles from my hillside, on an August day, he came upon a robber fly that had just captured a smaller insect. Apparently the fly was not hungry. Instead of dining immediately on its prey, it dropped it and pounced on it. It moved away and darted upon it again. It resembled, more than anything else, a cat playing with a mouse. Otto Emil Plath, in his *Bumblebees and Their Ways*, tells how the males of the species *Bombus separatus* fly about in the sunshine, apparently playing together for minutes at a time. And Miss A. M. Fielde, during ant studies that occupied the early years of the present century, noticed that worker ants would sometimes tussle for no apparent reason as though they were playfully wrestling.

As for myself, on several occasions, I have seen actions that suggested insects at play. But I could never be sure. The habits of these small creatures are so foreign to our own, their motivations so obscure, that I could never be positive my interpretation was correct. I stared at the robber fly and the robber fly stared at me. I thought: If I could only *be* a robber fly, for just a moment, how much I would know that is now speculation and conjecture. Of one thing I felt sure. The fly that stared at me had no desire to change places. He had no wish to be me for even a moment. He was content to be a fly.

AUGUST • 8

LEECHES At dawn this morning, I come upon a mud turtle, *Kinosternon substratum*, plodding along the path at the edge of the swamp stream. I pick it up and discover that it carries with it three glistening black leeches. One is attached to the underside of the shell, one to the side of the turtle's body and a third to a leg. Nothing in nature is entirely free from attack. Even the lion is surrounded by flies. These turtle-leeches, among the most lowly of attackers, have their own instinctive cunning. Oftentimes, they attach themselves in the "armpits" of the turtle's legs where they cannot be removed. If mud turtles should begin co-operating, scratching the leeches from beneath the legs of each other, nature's balance would be upset. That small show of intelligence would load the dice in the turtles' favor. All a species needs is a slight advantage. Multiplied by the compound interest of evolution, the smallest advantage soon grows great.

AUGUST • 9

THE PRIMROSE MOTH Not far from the straggling cluster of sunflowers at my Insect Garden, a stalk of primrose rises almost as high as my head. It is the lone representative of its species on the slope. There, in the early dawn today, I find two pinkish little moths with white hairs massed behind their heads. Both are in the same position, motionless, their heads thrust deep into the yellow flowers. When first I found such a moth in such a place, I thought it was dead or was being held by some unseen crab spider lurking at the bottom of the bloom. Now I know it was merely fast asleep, already bedded down for the day. For this is the habit of the primrose moth. It sleeps in a yellow bed during the August days.

GOLDEN ASTERS Across the swamp and the hillside, as summer advances, new flowers appear in their appointed time. The shaggy white blooms of the American great burnet now hang heavy with dew at the foot of the hill where the purple of Joe-pye-weed is beginning to appear. The great butter-yellow disks of the sunflowers shine at the tops of stalks that rise higher than my head. And north of the apple trees, like an island in a sea of open sand, the clump of golden asters is now in full bloom. For a decade and a half, I have seen it there. Only here do the golden asters grow; nowhere else along the swamp edge or up the hillside do I find them. Here they are rooted and here, slowly expanding their clump with the years, they seem as stable, as enduring as the trees around them.

AUGUST · 10

A BEWILDERED WASP About three o'clock on this windy afternoon, I stand once more beside the island of golden asters. In the sand of the open space, a *Chlorion* digger wasp is excavating her tunnel. Time after time, her orange-banded body disappears into the ground to reappear again, backing out, dragging a fresh load of dirt between her forelegs. At last, the burrow is finished. The wasp darts away, tacking low into the wind, searching the grass clumps for the black-striped green grasshoppers with which she will stock her tunnel.

Fifteen minutes go by. The wind pounds the trees. Swallows labor overhead into the gusts or shoot past riding the wind. Then the wasp reappears. She is coming on a straight line for the open space, flying heavily, carrying an inert green grasshopper like a long pontoon beneath her body. Leaving her paralyzed prey lying on the sand, heading toward the mouth of the burrow and half an inch from it, she investigates inside. Satisfied that all is well, she emerges, clamps her jaws on one of the grasshopper's antennae and swiftly drags it out of sight. Minutes pass. Then the face of the wasp appears in the mouth of the tunnel. She pauses briefly, then shoots away on a hunt for another grasshopper.

This time she is gone for twenty minutes. And while she is away I lay a wooden match across the opening of her burrow. What will she do, pick it up or push past it? She does neither. She turns over the open space, seems bewildered, lands, appears searching for the burrow that is almost under her nose. I watch her in amazement. The presence of a matchstick laid across the opening of the half-inch hole of the burrow has changed its appearance sufficiently to confuse her. She flies this way and that, as though taking bearings. She sits down on the asters, as if considering. Hers is an astonishing performance. Some time elapses before she finally locates the tunnel. Then, kicking with her hind legs as though spreading dirt, she knocks the match away. In her actions, she has revealed the keenness and the blindness of instinct. Apparently, so exact is the image of the entrance she carries in her memory that any change, even the slightest change, while she is away will confuse her on her return.

AUGUST · 11

PASSING THOUGHT We want to be cats with good homes. We want to make our living not by work but just by *being*.

AUGUST · 12

SQUIRREL RIDE A baby squirrel rode across the yard this afternoon. Just as I came out of the back door, the redoubtable Chippy was transporting one of her litter from the hollow silver maple where it was born to another tree. Unlike the carried kitten, it was not held by the scruff of the neck. Instead the young squirrel was curled around its mother's neck, back down, tail tightly wrapped over the top. It looked like a fur neckpiece. In this position, apparently gripped by the loose skin of its underside, it was being transported on its across-the-yard journey.

THE SPARROW FAMILY In the sunset tonight, I sat on the ancient plank, silvered by sea winds, that forms a bridge across a tidal creek far out on the salt meadows. There I ate my supper of sandwiches. And there, as I sat motionless, legs outstretched, leaning back against one of the upright posts, a family of sharp-tailed sparrows came trooping toward me, hopping and flitting over the tangles of cordgrass. They seemed to consider me part of the bridge. For a wonderful quarter of an hour, I had them all around me. They sang from the posts of the tumble-down bridge. They chased each other like scurrying mice. They hopped at times within two feet of my outstretched legs. With a wash of yellow along their sides, with ocher markings on their heads, the sharp-tails are far more handsome birds than the somber seaside sparrows. Never had I seen them to better advantage. In the sunset light, a kind of glow enhanced their modest beauty. In spite of gathering mosquitoes and numbing legs, I remained unmoving as long as they stayed. Walking home through the twilight afterward, I thought I would remember as long as I lived that time in the sunset, that family of sparrows unafraid.

AUGUST • 13

DANCE OF THE CADDIS FLIES To the Insect Garden late in the afternoon. A day of sunshine and glorious, high-piled cumulus clouds. A kingbird balances on the dead, topmost twig of a swampside maple, and russet dragonflies speed in wild gyrations among the apple trees. Two alight, facing each other, clinging to the same leaf. I shake the branch and their support lashes more and more wildly about. But the two insects are undisturbed. Thus they would sleep, anchored fast, through the windy nights.

Shining, shoebutton black, wild cherries hang in ripe clusters now where, but a few weeks ago, the trees were white with the foam of the flowers. As I walk along the garden paths, I eat bunches of the fruit with its bitter-tonic juice. At one tree, I watch a *Polistes* wasp alight on cherry after cherry until it finds one that has split open in its ripeness. Here it remains for minutes, motionless. It, too, is drinking the summer tonic of the juice.

Below the hillside, I find half a dozen dragonflies laying eggs, dipping their tails in the brown water of the swamp stream at intervals of several feet. They seem to be playing leapfrog with invisible companions. And all across the surface of the stream, among the drooping sweet flags, beside the path where I stand, dark caddis flies are dancing. I see them fluttering, bobbing about in the air, their numbers doubled by their reflections, by images as real as their rising and falling bodies, moving up and down above the stream's dark mirror. Here, too, are mirrored the clouds, tinted with sundown lights, and the caddis flies seem to be dancing amid them.

Like the dragonflies, the females will deposit their eggs in the water. From them will hatch those little case-makers of the stream bottom, the caddis fly larvae, clothing themselves in odds and ends to protect and camouflage their nakedness. All of the dancing insects remain within four or five inches of the surface, lifting and dropping, moving from side to side. Like the shimmer of heat waves across a brown plain, their dance goes on and on. The sunset fades. The twilight comes and still the insects are unwearied. At last I turn away, walk home and pack. Tomorrow at dawn, Nellie and I drive north for vacation days at a remote little lake, close to the Canadian line in the woods of Maine.

AUGUST · 14

THE ROAD Sunrise, today, found us on the road, beyond the Whitestone Bridge, following the Merritt Parkway to New England. Past red tobacco barns, up the Connecticut River Valley, across the width of Massachusetts, among the wooded hills of New Hampshire we rode. At dusk, this evening, we reached Chocorua. Beautiful country streamed away behind us all day long but we dared not stop. We were making distance. An automobile carries you far. But it limits your freedom. It binds you to the highway. You can't go across-lots and across-lots is where there is most of interest to see.

AUGUST · 15

THE LAKE Sunrise again. And we are driving north again. Not far from Conway, we pull off the road where birches rise in clumps beside a stone wall. Beautiful moths, white of wing with dark, wavy lines that make them blend with their background of birch bark, are laying masses of eggs and covering them with the tan, feltlike material of their body hairs. These are the notorious gypsy moths, the scourge of New England trees, introduced by accident eighty years ago and still unconquered after a battle costing millions.

For a time, we talk of swinging widely to the west, through Franconia Notch, to see once more the bridge where we stood watching the sunset die beyond the evening mist on that last day of our long trip north with the spring. But the sky is dull and hanging low and we drive on through Pinkham Notch. Rain in the notch, clearing skies at Gorham and the sun once more on the Shelburne birches. We ride into Maine, through Rumford and Skowhegan and Solon, north on the Arnold Trail, past the Bingham Post Office where, years ago, one August day, I found the galley proofs of *Grassroot Jungles* waiting, past Wyman Lake and Carney Brook and the dim trail Nellie and I once followed into the forest to live in a trapper's cabin.

And so to Jackman and beyond. Only a few miles from the Canadian line, we swing to the left, onto a side road that winds and twists, climbs and dips, crosses on planks a tumbling brook and reaches at last the wooded shore of a lonely lake. From the door of our log cabin, the lake stretches away, rimmed by dark forests and reflecting all the colors of the sunset ebbing behind the black hills of the farther shore. Here is the smell of wood smoke and balsam. Here is the wild laughter of loons in the twilight. Here is the silence of the forest night. And here is deep and dreamless slumber.

AUGUST · 16

THE CANOE A canoe returns some of the buoyancy to life. For hours, on this first of our forest days, we drift and paddle. We listen to the wind among the trees. We ride the ripples, breathe deep the scented forest air, pause to watch water-striders skittering away over the surface film. We explore little coves that end in half moons of driftwood, tangles of stumps and logs, silvery white, smoothed by the water and polished by the wind. We have no more serious business today than watching the clouds drift by. And thus, idly, we follow the circuit of the shore.

WILDERNESS WARBLERS Around noon, today, the woods about our cabin is filled with troops of migrating warblers, some adult, some immature. We see black-throated greens, black-throated blues, myrtles, magnolias, Cape Mays, redstarts, Canadas and Blackburnians. Between eleven and two, three little waves go through—one at eleven, one a little after noon and the third a bit before two. The warblers flit from tree to tree, hunting for caterpillars and small insects. Trooping along with them are chickadees and juncos, kinglets and red-breasted nuthatches. For the wilderness warblers, nesting time is over, migration time is here. They are feeding heavily, preparing for the next long flight to the south. So soon! How few the weeks since we watched northbound warblers by Milburn Pond! Now the tide has turned. All across the north, the waves of another migration are building up.

THE STRANGER It is the visitor, the traveler, the one passing through, who sees the beauty of the land unalloyed. The inhabitant looks around him through an obscuring cloud of responsibilities, worries and concerns. He sees his bit of the globe tinged by his circumstances. Thus, often-times, he sees it less clearly than does the transient who has cast no anchors there.

AUGUST • 17

RED MOON We awoke, about three o'clock this morning, to the calling of the loons—not their wild, echoing laughter but a long, muted, musical note. Looking out over the misty lake, we saw a red moon hanging low above the silhouetted hills and, amid trailing films of mist, a red moon and red ripple glints reflected on the surface of the water. Silently, we gazed at this strange and beautiful apparition glowing in the sky and in the lake. And, all the while, the loons sounded their faraway, lonely horns in the night.

THE FOREST TRAIL For a long time, today, we follow a green-carpeted forest trail. We steal soundlessly on living plush through moist and hoary woods. Moss spreads away in rich greens over the moldering trunks of fallen trees. Here and there, we come upon brilliant splotches of color where patches of slime mold are spreading amid the dampness and decay. At one point, reindeer moss runs like gray frost over the ground beside the trail. Toadstools, coral fungi, ghostly clumps of Indian pipes rise from the rich humus. Everywhere we are surrounded by the transient beauty of decay.

As we round one turn in the trail, we come suddenly upon a ruffed grouse family. The young birds, fist-sized, scoot away among fallen branches to one side of the path while the mother grouse flops and flutters at the other side to hold our attention. Then she darts across the trail and into the woods, zigzagging swiftly to gain the protection of successive tree trunks.

OLD EVENTS Sad are the days when the glory is gone from the earth, when we look about us with tired eyes, dimmed by the sediment of fatigue, and find we no longer can see beauty or wonder or drama in commonplace scenes. The most interesting things in the world to me are the old, familiar, recurring events of the earth.

WHITETHROAT A lone whitethroat sings in the woods back of the cabin each evening. This song is the only one of its kind we hear anywhere around the lake. The whitethroats have nested, the young are flying, the singing season is past.

AUGUST • 18

RAIN IN THE FOREST A loon laughed in the night and its laugh came back in fainter and fainter echoes from the surrounding ridges. Some time after midnight, the first raindrops spattered on the cabin roof and soon the small sounds of individual drops merged into the one greater sound of a drumming downpour. By eight this morning, the storm had become a drizzle and, by ten, only the irregular drops from the wet trees sounded on the cabin roof. Mist lifted slowly from the lake, clinging among the forest trees. A pair of loons, that had drifted inshore in the vapor, paddled away to the center of the pond. Fog-bound barn swallows appeared, sweeping low over the water and above the loons, coursing back and forth beneath the gray ceiling of the mist. All day, the clouds pressed low and rain came and went. The sun set, shrouded and invisible, behind heavy skies. Once when the rain seemed over, we walked a mile or so down the road, enjoying the brilliant greens of the wet mosses and the lacquered red of the bunchberries. Small, dark, northern bumblebees had taken refuge under the goldenrod leaves. They were unmoving and seemed bedded down for the night. Pale green and silver-spotted, a beautiful Luna larva fed on the leaves of a maple sapling at the side of the road.

Behind us we heard the loons leave the lake and fly away, their intermittent calling growing smaller and smaller until it faded away. Then, in almost perfect silence, with even the dripping of the leaves temporarily stilled, a branch, only a hundred feet away in the forest, broke with a crack like the report of a rifle. Why, in all the hours of its existence, had it given way at that particular moment? There was no breath of breeze. But there was the weight of water on it.

AUGUST · 19

CEDAR WAXWINGS Lying in the sun on a great shelving rock that slants down to the water's edge, we spend an hour and more this afternoon watching cedar waxwings. Their favorite perch is the dead top of a high spruce. One waxwing sits at the very tip of the topmost twig. As we watch it through our glasses, we notice a fly circling around it. The bird's head follows the movements of the fly like a kitten following a moving string. Several times we see waxwings swoop in long toboggans toward the water and flutter back to the tree top with large, dark-colored mayflies in their bills. A few of the *Ephemera* drift past us as we push off the canoe. We see, where it is riding the little waves below the shelving rock, a flat piece of wood around which fishline was originally wound. It now forms a raft to which clings the brown, translucent nymphal shell of a transformed mayfly.

We are halfway across the lake when three of the waxwings overtake us. They circle the canoe twice as though in a gesture of bon voyage. Then they return to their spruce tree. With their cheerful trilling, their peaceful ways, their friendly habit of passing food from one to another, the cedar waxwings have a gentle charm unexceeded by any other bird. I notice that, in flying over the lake, these birds often climb steeply, hang in the air and then point downward like an airplane that stalls and drops away in a nosedive.

ODD BELIEFS I have just heard three odd beliefs about wild creatures held by credulous natives of this north country. One is that loons can never go on land, that they lay their eggs in the water, as near shore as possible, and then push them up on dry ground to hatch! Another is that it is impossible for a swallow to land on a flat surface. And the third is that the beavers that begin a dam are given three chances to complete a good one. If it washes out a third time, they are dismissed and a whole new crew comes in to take their place!

AUGUST • 20

BEETLE GRUBS We were walking along the mossy forest trail, in the stillness of noon today when we heard a curious grating or rasping "zip-zip-zip" repeated at intervals. The sound came from a dead spruce forty or fifty feet high. All around the base of the tree were little piles of light-colored sawdust, the particles elongated in form. Five round holes, each about an inch in diameter, were visible in the trunk of the dead tree. The highest was more than thirty feet above the ground. From all the lower limbs hung strands of sawdust, in places so thick it resembled usnea lichen. The sound we heard, in the hot stillness of this August noon, was produced by the sawdust-makers, the big grubs of longicorn beetles. So loud was the rasping that we could hear it clearly when we had moved away twenty-five feet and more down the trail.

CHICKAREE Per ounce and inch, no living thing can develop more excitement than a red squirrel. We came upon a chickaree this afternoon sitting on a mossy log with a cone in its mouth. So swiftly did he tear it apart in getting at the seeds that, at times, he seemed surrounded by flying fragments. He suddenly saw us and shot up a tree trunk with his long, unwinding call, a sound like a notched spool whirled against a windowpane. Then he stopped on a lower limb. His indignation mounted. He barked, coughed, screeched, sneezed and spit until he seemed a tiny motor running away and about to explode. His tail flipped, his head jerked, he darted from one side of the trunk to the other, from limb to limb in the balsam tree. I sat down on a hollow log to watch him. The shell of the log collapsed and the squirrel jeered. His head popped out, first on one side, then the other of the tree trunk, his jowels plastered down and black-looking from the pitch of the cones. He seemed on the verge of apoplexy and fully five minutes went by before his excitement subsided. All along the moss-covered log, where we had first seen him, there were little piles of purple-fringed cone scales. During the August days of other years in the long, slow moldering-away of this tree trunk, how many generations of red squirrels had left there similar piles of cone scales! For who knows how long, the fallen tree had been the green-covered table of the chickarees.

AUGUST • 21

CORDWOOD All over the northern part of the country these days men are cutting cords of firewood for winter. We too are gathering wood, in a way, cording memories that will warm us later on.

THE VALIANT BUTTERFLY For half a mile today as I paddle close to the shore, an anglewing butterfly with silver commas beneath its hind wings—*Polygonia progne*, the Gray Comma—rides with us. It alights without fear on my hand, the one that is gripping the top of the paddle. With proboscis uncoiled, it dabs at my skin. I shake it off half a dozen times. It merely flutters up and settles down again. The wind strikes it and, at times, forces its wings far over, sometimes almost flat on my hand. For fifteen minutes, the valiant butterfly rides along, rising and falling and turning with strokes of the paddle. It clings there as it would cling to a flower tossing in the wind. Sometimes it holds fast to the back of my hand, sometimes it rides on the upper side of my thumb. I can recall only one other time when I have had a butterfly ride with me so long. That was many years ago, in the long corridor of a blowout in the Indiana dunes. As I walked slowly along, that day, a little copper fluttered down and rode on my shoulder.

A NEED Our minds, as well as our bodies, have need of the out-of-doors. Our spirits, too, need simple things, elemental things, the sun and wind and rain, moonlight and starlight, sunrise and mist and mossy forest trails, the perfumes of dawn and the smell of fresh-turned earth and the ancient music of wind among the trees.

AUGUST • 22

BEAVER BOG With a lunch packed in the canoe, we paddle the length of the lake this morning. At the far end, we come to a bay where a vast tangle of weathered driftwood is slowly sinking, through the years, into the black paste of the silty bottom. Pouring into the bay is a small stream of amber-colored water. This is the overflow from a beaver bog far back in the forest.

In places, this stream is hardly wider than our canoe. As we push up it, crimson-bodied dragonflies patrol ahead of us under the overarching alders and water-striders drift or kick in sunny places on the sluggish flow. Once, a small frog, its eyes encircled with a yellow line so it seems wearing gold-rimmed spectacles, leaps into the water, swims downward and disappears in a cloud of silt. A hummingbird whirs past, down the brook and back again.

At the end of half a mile, we come out into a wide sphagnum meadow, two acres or more in extent, bisected by the stream and edged with the feathery green of larch trees. At any moment, in this still and magic place, we expect to catch sight of deer or the towering form of a moose among the larches. In truth, only two days ago, a bull moose was seen standing at this very spot. We gaze about us at the reddish rosettes of the sundew, at the snow-white masses lifted by the tall stems of the cotton grass, at the red-veined pitcher plants, *Sarracenia purpurea*, leaning out over the brown water, at the aquatic plants in the flowing water beneath us, waving and fluffed like the tail of a frightened cat. Suddenly the silence is ended by a loud and raucous croak. We look up. A raven is flapping and soaring across the sky, heading toward the bare cliffs of Slidedown Mountain that rise like palisades to the north.

Beyond the sphagnum meadow, the alders close in again. Three times we drag our canoe around small beaver dams as we work upstream. It is nearing noon when we emerge among dead, upright, weathered trees into the deeper water of the beaver pond. An American bittern leaps up with a great flapping of wings. Green darner dragonflies, *Anax junius*, course over the pond and here, as in the brown water of my swamp stream at home, the females are dipping their tails and laying eggs under the hazy heat of noon.

We paddle first in one direction, then in another. We push into shallow water. We examine the work of the beavers. We watch kingbirds flutter out, hover, sweep back to favorite perches at the top of weathered stubs. Once, as we drift silently around a point of land, we are startled by a tremendous snort and a buck, feeding in the shallows, turns and leaps high in the air. It gains the forest in a series of jumps, a geyser exploding upward at each descent in the water. On a quiet night, that snort would carry for a quarter of a mile or more. It is as good an alarm note as the slap of a beaver's tail.

We eat our lunch in a Valhalla of Fallen Giants, a secluded place where great tree trunks, some seventy feet long, lie robed in moss. A whiskey jack, the white-headed Canada jay, flies toward us across the water. Here we are in the heart of the great northern woods. Hardly a mile away lies the Canadian boundary. And beyond stretches the forest, on and on, without roads, without houses, a land of scattered lakes and few streams, an area where a man could stay lost for days or weeks or forever. As we eat, we hear, high above the balsam trees, the drone of an airliner flying from Boston to Quebec. This tenuous link binds us to the twentieth century. All other sounds around us—the call of kingbird and whiskey jack, the breeze among the balsams, the rattle of dragonfly wings, the snort of deer—all these are ageless sounds that give no inkling of whether we are sitting on this moss-covered log in the mid-twentieth century, the eighteenth, sixteenth or tenth.

Paddling slowly home in the late afternoon, we pass one of the lake coves, with its half moon of driftwood. A black shadow runs among the silvered timbers. It is a mink sinuously twining in and out, running at top speed, following the trail of some quarry. It moves with a fluid, forward motion, like a blacksnake among branches. I strike the side of the canoe with my paddle and wake the echoes in the surrounding forest. In a flash, the mink disappears. Then its head pops above the root of a weathered stump. It peers at us intently with glinting little eyes. Then eyes, head and mink slip swiftly out of sight.

AUGUST · 23

THE SNOWSHOE RABBIT At six this morning, the sun, rising behind our cabin, throws the spire shadows of the spruces half a mile out across the misty waters of the lake. Under one spruce, where we see it each morning and evening, a snowshoe rabbit is feeding in a little patch of grass. Big-footed and brown, it hops leisurely about. In only a few weeks, the brown will be gone and its fur will be as white as the drifts among which it will spend the long northern winter.

JEWELWEED AND HUMMINGBIRD There is a shallow valley or low-land stretch a hundred yards back of the cabin. It is filled, from end to end, with flower-laden jewelweed. This is the province of the hummingbirds. They dart about us in concern whenever we walk that way. This afternoon, as we skirted the jewelweed about sundown, two ruby-throats came streaking toward us. They zoomed over the trees behind us, one bird uttering a sharp squeak as it went by. Later, one of the hummingbirds swept back and forth over the area repeating at intervals of a second or less a series of grating sounds, a kind of "De-Deet!" or "Dit-Durr!" This double note was repeated twenty times or more. The bird landed and was still, took off and began calling again. A little farther along the jewelweed, we discovered a young hummingbird perching on a twig. It lacked the metallic brilliance and the sleek appearance of the adults. While we watched, it whirred into the air and darted down to feed among the yellow blooms below. One hummingbird Nellie saw today hovered for half a minute over a clump of pearly everlasting, probably finding small insects or spiders there.

AUGUST · 24

THE DESERTED LUMBER CAMP Again we push our canoe up the amber-colored stream, under the alders, across the sphagnum opening, over the succession of small dams and, at last, into the dark, slack water of the beaver pond. Today we hear a nuthatch deep in the forest and the mighty hammering of a pileated woodpecker. A white cabbage butterfly, strangely out of place in this wilderness, drifts among the gaunt stubs of the dead trees. As I paddle to the far end of the pond, the sunshine reflects from the running ripples into my eyes with a steady blinking light.

We beach the canoe and explore a trail that leads into the woods, a path no doubt originally laid out by deer. The breeze is rising and the sounds of the forest increase. Walking silently on moss, we hear long creaking moans and sometimes, when the wind is stronger, a note like a trumpet blown far away. Suddenly the trail opens out into a clearing and we find ourselves among the low, weathered buildings of an abandoned lumber camp. All across the opening are the magenta blooms of the fireweed, that remarkable plant so swift to draw its veil across the forest scars. At the top of each flower pyramid there are buds, while at the bottom the silken seeds are already formed. As gusts rake the plants, the fireweed silk rises in white clouds, like puffs of glistening steam. Then, higher in the air, the silk of the seeds separates until the whole resembles a multitude of flying insects.

The heat of the day, a sunny, grasshoppery day, presses down on the clearing. Wherever there is soft earth, we see the tracks of deer. Curiously, when a portion of the forest is cut, it tends to attract deer rather than frighten them away. They come to feed on a special delicacy, the lichens of the felled tree tops.

WHITE-WINGED CROSSBILLS We are sitting by the canoe, eating a lunch of sandwiches with handfuls of fresh-picked blackberries for dessert, when the air around us is filled with dry, grating calls. Half a dozen birds, dark and sparrow-sized, sweep over us to alight among the tangled driftwood and weathered stubs beyond. Theirs is a call that has mystified us for days. Once we heard it on one of the coves of the lake, once among the alders, once near the jewelweed. But we were never able to trace it to its source.

We swing our glasses to our eyes, checking the field marks of the birds. Some, the males, are rosy red; the females are greenish gray. All have white wing bars and bills with mandibles crossed. We have never seen this species before; but identification is positive. We add the white-winged crossbill to our life-list. These birds tend to wander, to be in a region one year and not the next. On none of our previous trips to Maine had we encountered them.

One of the males mounts to the tip of a stub and gives a long, sweet song, suggesting that of a goldfinch but with more rounded and musical notes. For more than five minutes, the crossbills hop and flutter about the driftwood. We see them do the same thing over and over. Alighting on a weathered stub, where a barkless branch joins the trunk, they lean far down one side and peer or peck underneath. Then they lean down from the other side and repeat the process. At times they pause and swallow something. What are they doing? Are they getting small spiders or insects? We examine similar sites but can see nothing of interest. However, we are not crossbills. When the birds dart away with their chattering cries, they leave with us this minor riddle still unsolved.

Over the pond, as we are pushing off, a goshawk rises up and up into the wind. It mounts in quick circles. Once it swerves and plunges fifty feet seemingly in the wink of an eye. What speed, what life, what keyed-up nervous tension its aerial maneuvers display! It is a sight to stir the sluggish blood of any earthbound creature!

AUGUST · 25

THE OLD FOREST ROAD A winding road and a winding river tax all my powers of decision. Each new turn promises a fresh surprise. Just one more turn! Let's see what's around just one more bend! And so from twist to turn, from bend to curve, I go on and on.

I remember my failing when we start down an old forest road this morning and set a time when we should turn back. This moss-covered road, long abandoned, skirts the south shore of the lake on its winding way to the deserted lumber camp. It is a lonely road where few humans go. We see deer tracks of different sizes imprinted in the green and russet moss. In one muddy place, a fawn track stands out as small and delicate as a shell. And once, as we mount a little ridge and descend again, we see where a moose has followed the road before us.

We wander slowly from turn to turn among the enchantments of this forest road. We stop to look at mottled toadstools. We examine interrupted ferns. We turn aside to admire a trillium with its bright red, persimmon-shaped berry at the top. We pause to see masses of magenta fireweed and patches of shamrock-green wood sorrel. Where, for a dozen yards, bushes are red with fruit, we harvest wild raspberries, eating them as we go.

In boggy dips, frogs leap away into the grass. In sunny glades, locusts hover on crackling wings. We come upon jungles of fallen tree tops inhabited by winter wrens and white-throated sparrows. And, along the way, we meet little families of birds feeding in the forest—a family of juncos, a brood of ruffed grouse, a family of golden-crowned kinglets hopping about amid the dead branches of a balsam.

At the final bend, before we turn back, we eat our lunch. We sit on a fallen birch. The partially decayed wood within the bark forms a soft and mosslike cushion. At several places, as we slowly return, we stop to look at maples already turning red. Last night the thermometer dropped to thirty-six degrees. Today there is a chill in the deeply shaded woods, even at noon. On this north country road, there is little of the August heat that lies on my Insect Garden at home. For here we are half a thousand miles north of Milburn Pond. Here we are in a land of earlier autumn.

AUGUST · 26

MOONLIGHT AND MIGRATION After dark, on this last of our forest nights, we paddle into the lake, floating under the starry sky, and watch the moon, still almost full, rise above the ebony spruces. Far down the lake, for a time, the loons call softly in the moonlight, a mellow, sad, lost and lonely sound among the mountains. Then all is still and I sit enveloped by memories of another night, years ago, on Saranac Lake in the Adirondacks, with my son, David. It, too, was a night filled with stars.

Then, down from the silent sky, come frail little wisps of sound, tiny cheeps and chirps, the calls of small birds migrating. Long after we pull the canoe up on the shore, these almost inaudible calls continue. We listen, standing there beside the moonlit lake with the Great Dipper hanging low above the silhouette of a birch tree. We speak of the power of this urge that sends a parade of little birds across the heavens. We think of the immense distances that lie ahead of them. And, before we go into the cabin and shut the door, we talk again, as we have talked a hundred times, of the mystery of migration, here symbolized by tiny voices almost lost in the vastness of the night.

AUGUST • 27

FOG AND SPIDER WEBS In dense fog, we drive away at seven in the morning. All the little spruces, on the way to the Arnold Trail, are decorated with moisture-bejeweled spider webs, webs that hang like the metal dishes of an apothecary's scales or like saucers supported by strings rising to an apex. They are the work of *Frontinella communis*, the bowl or doily spider. Sometimes as many as a dozen webs festoon one small tree. All the web-makers cling motionless beneath their shallow bowls of silk.

Heading south, we drive all day through signs of coming fall. Migrating martins and barn swallows are thick above the Kennebec. Concentrations of grackles drift across the open fields. And, toward evening, we see the late-summer flocks of the starlings turning and whirling and weaving like black ballet dancers performing on the stage of the sky.

AUGUST • 28

A DOG HOWLS AT 8 A.M. On our way home today, we drive through Concord to see, once more, the rivers, Fair Haven Bay, Walden Pond and Ministerial Swamp. We have lunch at Helen's Restaurant. The proprietor tells us of a dog that howls on the main street each morning at exactly eight A.M. A whistle, sounding at that hour, probably hurts his ears and sets him off. Each time, he howls for a minute or more.

As we leave Concord—out past Mill Brook and Emerson's home

and the site of Thoreau's bean field—I recall a laborer I met here during the jobless days of the great depression. He declared: "I'm like to starve to death in this historic town!" But he said it wryly, without bitterness, implying that if he had to starve to death anywhere he was glad it was going to be in Concord.

Home at 7:45 P.M.

AUGUST · 29

DRAGONFLIES AS WEATHER INDICATORS To the Insect Garden in midmorning. Change is everywhere. I see the brown edging of the sunflower leaves, the yellow grass of summer's latter days, the trampled appearance of the drooping sweet flag leaves. The marsh is splotched with the purple of Joe-pye weed and white waves of climbing boneset run among the cattails. The larger dragonflies are coursing back and forth high above the hillside. Like swallows, they are weather prophets, or at least weather indicators. On dull, heavy days of imminent storm, both swallows and dragonflies hug the earth; on the clearest days, they ascend. The reason for both is the same. The small insects, upon which they feed, fly higher on the clear days.

WIND WRITING In the afternoon, the wind rises. It blows from the south, in long gusts, beneath a clear, burnished sky. I see tree swallows hanging in the air above me or whirled away downwind. Looking out over the swamp, I watch the cattails waving unevenly, their rise and fall a kind of wind writing. I can trace the progress of a gust moving down the length of the lowland. Its advance is shown by the bending of thousands and thousands of slender leaves, by the nodding of upright cattail stems, by the shimmer of wild cherry leaves, by the bending of the plumed phragmites and by the bobbing of the Joe-pye weed and the boneset.

AUGUST · 30

THE FRIENDLY PIGEON Once in a while, for some unaccountable reason, a wild creature will develop an attachment for some particular human being. I remember reading of a remarkable instance of the kind that occurred at the Amsterdam Zoo in Holland. A caged bittern there preferred the company of the director of the zoo to any of its own kind. It even drove its mate from the nest and tried to induce the director to sit on the eggs!

Some years ago, there occurred near here a somewhat similar attachment of a bird for a human. A friend of ours was in the habit of scattering grain on the lawn near French windows looking out on a terrace. It attracted thrushes, wild sparrows and mourning doves. Soon she noticed that a blue-and-white pigeon was alighting among the other birds to share in the food. It was the only one of its kind to come.

A little later, the pigeon began following her about when she walked on the lawn. Then it commenced appearing at the French windows on the terrace at about five o'clock every evening. It would flutter down and alight near the sill. As soon as she opened one of the windows, the pigeon walked inside. Sometimes it would walk clear around the living room as though on a tour of inspection. At other times, it would fly directly to the top of the fireplace screen. There it would perch motionless, following the movements of the woman with its eyes.

This strange behavior continued for more than two months. Each time the bird would remain in the living room for an hour or so. Then it would return to the French windows. As soon as they swung open, it would fly away. Our friend never learned where the pigeon came from; she never discovered what its history was. But, evening after evening, it arrived as though visiting a friend. It came at about the same time each day, remained about the same length of time on each visit, and, just before dusk set in, it went on its way. It seemed to feel a special friendship for this one human being.

AUGUST · 31

AMBUSH BUGS Among the purple flower spires of the butterfly bushes these days, I find secreted the strange little ambush bugs. Hardly half an inch long, they hide themselves among the florets, with enlarged forelegs outward, and thus wait for some insect victim to alight within their reach. I see these little apple-green and dark brown predaceous bugs holding cabbage and grayling and red admiral butterflies, flower flies and even bumblebees. The red admiral has a wingspread of fully six times the length of its captor. Yet, when caught, it flaps only two or three times; then it becomes quiescent. Apparently, like the robber fly, the ambush bug injects an anesthetic through its sucking beak into the body of its victim.

I notice that, on the butterfly bush, as the purple spearheads of the blooms progressively die from base to tip, the ambush bugs move farther and farther out. They keep among the fresh flowers until the tip is reached, wasting no time among the dead, dry florets that will attract none of the nectar hunters.

I try to pick off one of the lurking bugs. Like a disturbed Japanese beetle, it drops to the ground, feigning death. I pick it up between thumb and forefinger. Then it discloses another of its defensive tricks. It gives off a penetrating, unpleasant, stink-bug smell. Ten minutes later, I can catch the sickish odor on my hand. I toss the feigning ambush bug high in the air. As it falls, it quits playing 'possum, opens its wings and sails across the garden with the breeze. Its flight is heavy, beetle-like, rather slow. It bumps to a landing among the wild cherry leaves. I see it slip from one leaf, tumble to another, lose its hold, slide from leaf to leaf and come to rest in a grass clump below the tree. Apparently this creature, so well fitted for clinging amid the flowers of goldenrod and buddleia bushes, is unable to hold onto the flat surface of the wild cherry leaf. I find that this is so. I retrieve the bug from the grass clump, place it on a cherry leaf, steady it there for half a minute to enable it to get a good hold, and then lower the leaf gently until it hangs in a natural position. The ambush bug slides off backward and ends in the grass clump again. I try it over and over. It is able to cling to the leaf as long as the leaf is held almost level. But as soon as it tilts down at a steeper angle than forty-five degrees, the bug slides off.

CHAPTER NINE

September

SEPTEMBER • 1

THE HOUR OF THE OWLS Now, the screech owl families are beginning to call back and forth from the darkened trees just before dawn. It is the hour of the owls. After the activity of the night, they seem to pause and engage in conversation before the sleep of the day. Last night, in moonlight and mist, one screech owl sat in the silver maple tree, uttering its quavering call at intervals for an hour or more. The sound, coming through the mist, had an added eerily disembodied quality.

A few days ago, when I walked across the yard at five in the morning, the sky was bright with stars. An owl, I think it was a barn owl, circled twice above me, over the trees and over the yard, silent as drifting thistledown. There was no flutter, no airy murmur of wings, no auditory warning for its prey. As far as sound was concerned, as it wheeled against the stars, it did not exist.

A few years ago, a minister brought me an unusual owl that had flown into his screen door. It was a young screech owl; but its plumage, instead of being gray, was rusty red. It represented the red phase that sometimes appears among these owls. The bird had been shaken up but it was not injured. I kept it overnight and released it the next day among the old trees of my Insect Garden, placing it in thick foliage where jays would be unlikely to find it.

Most of the time, during the evening and night it was with me, the owl perched at the top of my study door. We fed it grasshoppers, chopped beefsteak and three praying mantes. Grasping a mantis with one foot, it would hold it out and tear at its food with its beak. Once, it went over all the toes of its feet, biting and cleaning them one after the other.

When two pencils clicked together on my desk, it was alert in an instant. An airplane flew over the house and the owl froze motionless on its perch. Several times, when people passed under a street lamp across the street, it would crane its neck to see them. It seemed particularly interested in *distant* movement. I noticed with what rapidity the pupils of its eyes contracted and expanded with every change in light intensity. Once, when light from my desk lamp suddenly illuminated one side of its face, the pupil of the eye on that side instantly contracted and, for a time, remained hardly half the size of the pupil on the other side. Whenever it was disturbed, the bird would utter a low, fluttering call, a kind of whisper song, a miniature of the sound I now hear during the hour of the owls before dawn.

SEPTEMBER · 2

WATER Of all the drinks in the world, the finest to the healthy man is water. If all the springs ran Coca-Cola, water would be sold by the ounce.

SEPTEMBER · 3

THE IMAGINARY SQUIRREL As I was racing against time to finish a piece of writing today, my telephone rang. A woman I did not know asked me to come quickly. She had discovered something of spectacular natural history interest in her backyard, a mile away. What was it? She wanted to surprise me. I was very busy at the moment. Well, it was a petrified squirrel! A what? A squirrel that had become entangled in the nest of fall webworms and had died and dried up there. Was she . . . ? How could . . . ? I went. I should have known better. I did know better. But a writer always seems to welcome a chance to quit writing—even when the excuse is as wildly improbable as this one. The woman rushed me to the rear of her house. She led me to a tree. Fifteen feet above the ground a limb, sure enough, held a silken webworm nest. There in the nest—she pointed—was the brown mummy of the squirrel. There—correction please—was a small mass of brown leaves.

SEPTEMBER • 4

REDWING'S ROPE This is the third summer that a male redwing has used the same peculiar roosting place at night. A couple of miles from here, at the end of an inlet from the bay, there is a dock where a dozen motorboats are kept. Every evening—again this summer as in the previous two—this particular redwing alights on a coil of rope on the deck of the same boat, hops down inside, and sleeps there, protected by a little stockade of hemp.

THE FIVE STARLINGS A whole family of starlings, five of them, are anting at the same time today. Their odd positions, their clownish actions, their frenzied rubbing of bill on body, makes the scene of their activity resemble a five-ring circus in miniature. Once, this summer, I saw a speckled robin, not long out of the nest, engage in anting. Apparently it is instinctive rather than learned. The ants at the favored spot under our maple tree are recorded in the lists of myrmecology as *Formica fusca*. It is one of the commonest species, ranging over the whole north temperate portion of the globe. In captivity, *fusca* workers have lived as long as six years.

SEPTEMBER • 5

MANTIS VERSUS YELLOWJACKET This afternoon, about 3:30 P.M., I come upon a green praying mantis, *Tenodera sinensis*, clinging to the bloom of a butterfly bush at the Insect Garden. It is an adult and winged. In the grip of its spined forelegs it holds a Monarch butterfly, one of a number now drifting along the hillside, beginning the long flight of the fall migration. The mantis commences eating its prize. As I watch, a yellowjacket wasp alights on the butterfly. It, too, is hungry for insect meat. Immediately it falls to with its mandibles. It pays little attention to the predatory mantis, a dozen times its own length and infinitely greater in bulk. To get at choice morsels, it even walks over the head of the feeding mantis. At last, it manages to get a little ball of meat, about the size of its head, stuck to the face of the mantis. This is too much. The mantis rears back, lets go the butterfly, paws at its face like a dog bothered by a fly. The wasp pursues its meat. It alights on the head of the mantis, runs about, seizes the morsel and flies away. Then it returns to the butterfly that has remained wedged among the florets of the buddleia bloom. The mantis retreats six inches and stares at the wasp with the intentness of a Manx cat. But, perhaps because it is not especially hungry at the time, it makes no move to drive away the wasp or return to the feast. The little banded insect is undisputed master of the field.

I lift the butterfly by its wings from the buddleia bloom. The wasp clings in place. The butterfly slips from my fingers, falling four or five feet to the grass. The yellowjacket rides it all the way down. I pick up the butterfly again and, holding it at arm's length above my head, fully eight feet above the path, let it drop a second time. Again the wasp rides downward through the air. Only when the Monarch strikes the hard-packed path with a bounding thump, does the yellowjacket let go. It darts into the air, circles several times and settles down on the butterfly again. Perhaps the meat so desperately sought is for late-maturing larvae in the nest.

The relationship between mantis and yellowjacket is one that has interested me for years. Only late in the season do they come into conflict. I remember seeing one yellowjacket circle around a mantis and alight on its abdomen near the tail. The mantis reached around with one of its forelegs, clamped the spined trap it formed over the wasp, and calmly devoured its captive. At other times, late in fall, I have seen yellowjackets flying about the larger insects, more than once alighting near the tail. Perhaps they habitually approach the mantis from the rear. Late in September, one year, I came upon a male mantis lying among daisies, its forelegs moving feebly, its whole abdomen gone and the muscles of its thorax being consumed by half a dozen ravenous yellowjackets. The mantis, otherwise uninjured, appeared to have been killed by the wasps. At the end of the cannibal feast in which the female mantis often devours her suitor at the conclusion of mating, I sometimes find yellowjackets dining on bits of meat attached to the legs and wings left behind by the female. Without doubt, the banded wasps have a taste for mantis meat when they can get it.

SEPTEMBER • 6

DOG-TROT How much we miss when life has to be lived at a dog-trot!

SEPTEMBER • 7

THE DISAPPEARING BIRDS "Ever since 1928," a correspondent has written me from Pennsylvania, "I have had children in school and I have had them ask every teacher they have had the same question. No one could answer it. The question is: What becomes of the birds?

"The robins, starlings, sparrows, red-winged blackbirds are all plentiful in this area. Yet I almost never find a dead bird, not even a skeleton. What happens to all the birds? That is something that has puzzled me for years."

It is an amazing thing, considering the vast numbers of birds we see about us, how small is the number we ever encounter dead. We see a few along the highways killed by speeding cars. In the hunting season, we find a dead water-fowl now and then floating on the surface of the water. But, in the main, the birds just disappear. We see only those that are active and alive.

There are a number of reasons for this. In the first place, ailing birds tend to hide. They secrete themselves where they will not be disturbed or found by their enemies. Thus death most often comes to them in hiding places rather than in the open. Most birds spend their time among bushes or grass or in wooded areas. The chances of coming upon their small bodies in thick brush or among the tangles of grass in a field are always remote.

Moreover, birds that are sick or injured are the natural prey of many foes. Hawks, owls, weasels and foxes are ever on the alert for them. The birds least able to escape, least vigilant in avoiding danger, are the ones most often devoured. Many birds thus disappear even before disease can bring about their death.

Those that die in their hiding places are soon disposed of by scavengers of many kinds. In the economy of nature, their flesh feeds a host of smaller creatures. Carrion beetles alight at the spot. Flesh flies deposit their eggs and their larvae play their part in removing the bodies of the birds that die. Bacteria hasten decomposition. And ants quickly discover the feast.

Besides these small scavengers of the ground there are the scavengers of the sky. When larger birds die, vultures soar to the spot. Thus many factors, working together and producing their cumulative effect, provide the answer to the question: What becomes of the birds when they die?

SEPTEMBER • 8

DRAGONFLIES IN THE SUN All across the Insect Garden, on this hot September afternoon, I see the little, red-bodied dragonflies of autumn. They belong to the genus *Sympetrum* and are the last to come each year. They zigzag over the open spaces, cling to the apple leaves, alight on the sunflowers and the buddleia bushes. They are the latest in season of all the dragonflies I see. Every September it is the same. I become aware, almost with surprise, that the hillside is alive with their glinting, darting, colorful bodies.

Today, I notice an interesting thing about these creatures of the sun.

Watching them land, I see several, after they have come to rest on a twig or weed top, turn sharply around, some in a full half circle. Then I observe that most of the perched dragonflies are heading in the same direction. They cannot be heading into the wind for the air is still; not a breeze stirs along the hillside. The time is about three P.M. The sun shines from west and south of the zenith. Is the sun causing the dragonflies to change position when they land? I keep track of thirty-seven successive landings. Not one of the insects comes to final rest facing directly into the sun. If they land in that direction, they quickly shift position, turning so the full glare of the sunshine comes from the side or back.

This is something I have never noticed before. Perhaps it is a peculiarity of this particular kind of dragonfly. Is it really the sun that is making the insects turn after landing? I try an experiment. I break off a long, dry twig. I hold the upper end against the feet of a clinging dragonfly. It steps off onto the twig. I hold the twig upright and rotate it slowly until the clinging insect faces directly into the sun. Its great, bulging eyes receive the glare head-on. The dragonfly moves its head as though the light bothers it. Then it changes its position, swinging around in a quarter of a circle to the left. I slowly rotate the twig until the insect is facing the sun again. A second time it moves around to the left until it is sidewise instead of head-on to the glare. This it does over and over again. As often as I move it, it moves back. I should add that these *Sympetrum* dragonflies have an unusually calm disposition for such high-strung insects. Sometimes they alight on me when I am standing or sitting or moving slowly. In some species, the males are known to fall asleep in the sunshine at which times they can be picked off by hand. It is the males that are brilliantly red; the females are brown. The species I see at the garden is *Sympetrum vicinum*.

Several times, the dragonfly on the rotating twig takes wing. But each time, it cannot resist alighting on the tallest support at the spot where it is settling down and this is the twig which I hold aloft there when I see it preparing to land. For a while, I rotate the twig in a succession of coglike movements and the dragonfly performs as though in synchronization, turning away from the sun in a similar series of movements. Then I rotate the twig slowly but steadily. The insect keeps turning against the direction of movement like a man going down an escalator that is going up. Before the little dragonfly escapes from this bewitched twig, I turn it so it faces the sun a total of seventy-six times. It turns away from the sun seventy-six times. In addition to that, I watch thirty-two other *Sympetrum* dragonflies alight on various supports about the garden. Not one of them comes to final rest facing directly into the sun.

SEPTEMBER · 9

ONLY ONCE Labor Day is past; school children have returned to classrooms; summer is nearly gone. I wander along the swamp stream under the midmorning sun. Green castles of algae rise in a little bay of still water. A fish is swimming among them. As it catches sight of me, it shoots away with a powerful drive of its tail. Behind it the product of currents set up by its sudden rush, a ring of green scum half a foot in diameter, drifts upward. For all the world like a smoke ring, it turns as it rises, floating through water instead of air. For half a minute, the green ring drifts, rotating slowly. Then it disintegrates. I have never seen this before; I probably will never see it again. Only once, only once in a lifetime, am I likely to experience this odd little adventure with coincidence.

SEPTEMBER · 10

A BLUEJAY HUNTS ANGLEWORMS A singular instance of a bird adopting food and feeding habits foreign to its kind has just come to my attention. A young bluejay fell from its nest last spring and was raised by a boy of this neighborhood who was more familiar with feeding robins than bluejays. The bird developed a great fondness for angleworms. Later, after it had been released, it would run about the backyard in search of its favorite food. It would hunt the earthworms like a robin, cocking its head on one side as though listening, running over the ground, tugging to drag a reluctant worm from its hole. It is the first worm-hunting bluejay of my experience.

SEPTEMBER • 11

TIDES OF LIFE For days now, Monarch butterflies have been drifting through the Insect Garden, pausing to sip nectar at the buddleia bushes, then sailing away along the hillside to the south. The parade of the butterflies is in full swing. I see the Monarchs over the sea meadows, above the waters of the bay, in our backyard. There are tides of life as well as tides in the sea. The Monarchs, like the birds that pass by day and by night, are part of the great southflowing tide of the autumn migrants.

THE DUSTY MILLER The dust of the dusty miller may have a special life-saving function on occasions. Such moths seem to escape from spider webs, frequently, by leaving the powdery, shed wing scales on the sticky threads. Like the bird that escapes from the cat by leaving its tail feathers behind, like the daddy-longlegs that breaks away from a captor by discarding one or more of its legs, like the lizard that sheds its tail in wriggling free, the moth, on such occasions, saves its life by leaving part of its body behind.

SEPTEMBER • 12

THE PIGEON HAWK AND THE SWALLOWS For an hour or two, toward the end of this day, I walk over the sea meadow and beside the bay. The tide is out and fine lines of scum lie like spider silk on the wet, glistening mud. In the still air, I hear the squeaking, sizzling sounds of the fiddler crabs. Sea lavender is in bloom and I see it before me, spreading in delicate blue carpets, as I walk back across the moor. A laboring cicada, like a slow-flying hummingbird, passes me. Far from any tree, it bumps to a landing in a clump of tall cordgrass. Overhead, tree swallows plow back and forth, feeding on insects before congregating for the night in the dense phragmites. Nearer the phragmites, across the meadows, the migrants mill about in a living cloud.

As I watch through my glasses, a pigeon hawk drops like a thunderbolt through the cloud. The momentum of the hawk carries it up in a rocketing zoom and it strikes again, this time from below. The sky seems full of swallows; it appears impossible to miss. Yet the hawk climbs with empty talons to plunge again. It repeats its dive and zoom over and over and each time the birds, swirling and crisscrossing in the air, evade it. Then it picks out a single bird, follows it up and down, missing it by inches. After each miss, the swallow flutters upward, seeking to keep above the hawk. It chooses the air for a duel of maneuverability rather than seeking refuge in the phragmites. Five times it escapes slashes by the pursuing hawk. It is still safe, all the hundreds of swirling birds in the swallow cloud are still safe, when the little falcon gives up and flies away.

SEPTEMBER · 13

THE BAT About 3:30 P.M., on this hot and brilliant afternoon, I come upon a little red bat in a wild cherry tree at the garden. It is hanging upsidedown from a small branch on a level with my head. Its wings are wrapped tightly about its body. It resembles a reddish, furry cocoon hanging there. In spite of the brightness of the sun, it is sleeping soundly. I photograph it from several angles. Each time the shutter clicks, I notice the little bat gives a convulsive jerk. It is responding to the high-pitched metallic sound even in its sleep.

SWAMP WALK In the still heat, I follow the swamp walk. Yellowjackets are zigzagging over the grass everywhere. Their numbers are at a peak now. A hover fly hangs above the path, its blurring wings beside its body seeming almost solid so that the insect has the appearance of a bow tie suspended magically in the air. In the sun, it is honey colored. When I come upon a box tortoise, moving in slow motion down the trail, it pulls in head and feet and closes its shell. A field cricket, fleeing from my approaching feet, scrambles and slides up over the curving roof of the turtle's portable house. Now the sweet flags, the very symbol of mortality, lie flat and brown. The green forests of spring are the brown plains of fall. And up the slope of the garden, everywhere I look, the brown is increasing. The heads of the sunflowers are bending low, weighted with the swelling seeds. This is the end for which the earlier days were lived. There is an airy "swoosh" and three ruby-throats converge on the butterfly bush. Watching them through my glasses, I marvel at their swift co-ordination. They move ahead and back, rapidly obtaining nectar from the innumerable florets of the massed buddleia blooms. Their slender bills dart in and out of the purple flowers with a rapidity that brings to mind the moving needle of a sewing machine. They stay for only three or four minutes. Then they shoot away over the apple trees like little rockets. I walk home, exhilarated by their pulsing life, remembering those other hummingbirds among the jewelweed in the forest of Maine.

SEPTEMBER • 14

THE UNCLE A small boy in this vicinity has learned to recognize most of the birds he sees. One result is that he spends a good deal of his time debating identification with his uncle, a man of fancied knowledge and strong opinions. He recently told his uncle he had seen a hawk. His uncle said: "There ain't any hawks on Long Island!" He showed him a hawk. His uncle said: "That ain't a hawk. That's a sea gull!" The boy is young, the uncle old; so ignorance outweighs knowledge and age is given its due respect.

SEPTEMBER • 15

WHAT IS A SWALLOW'S NEST MADE OF? On the 26th of May, I set down in this journal a question about what the barn swallows were putting into the nest they were building under the wagon shed roof. Did it contain anything besides mud and grass? I was not sure. I noted: "Come back in September and I will tell you!" This is September. And I can tell you. For, today, with the swallows gone, I pried off the deserted nest, brought it home, tore it apart and analyzed its contents. The results surprised me greatly.

The nest, complete, weighs 9½ ounces. Of this, about 8 ounces consist of clay and mud. Lining the cup of the nest are 33 small white chicken feathers. They are all laid so their curve corresponds to the curve of the nest and together form a downy cushion. Beneath them is a mattress of wiry little roots, probably grass roots. I count them. In the mattress, I find, there are 114 rootlets and pieces of roots. In the mud of the nest, I find a total of 718 grass stems and fragments of grass. They range from half an inch in length to one, on the outside of the nest, eight inches long. No doubt many of the smaller fragments were broken up when I dismantled the nest so the original number of pieces of grass inserted by the swallows was less. In addition, there are 89 small twigs and leaf stems. Most of these are, apparently, the dry stems of maple leaves. One is recognizable as part of an arborvitae twig.

In the mud of the nest, I find various small hard objects. I wash them out and arrange them in five little piles. In the first pile, there are 34 fragments of coal. In the second, there are 18 bits of brick. In the third, I find 9 pieces of shells; in the fourth, 8 bits of coal ashes; in the fifth, 10 fragments of stone, mostly quartz. This makes a total of 79 different hard fragments.

What is a barn swallow's nest made of? A good deal more than just mud and grass. It is far from the simple cup often imagined. This nest, in its 9½ ounces, contains a total of more than 1,000 objects. A summary shows that, in constructing their nest, the barn swallows brought to the wagon shed 8 ounces of clay and mud, 33 chicken feathers, 114 rootlets, 718 grass stems and fragments, 89 leaf stems and pieces of twigs, 34 fragments of coal, 18 pieces of brick, 9 bits of shells, 8 minute lumps of coal ashes and 10 pieces of stone.

SEPTEMBER • 16

THE MEASURE OF AN ENTHUSIASM The measure of an enthusiasm must be taken between interesting events. It is between bites that the lukewarm angler loses heart. It is between birds that the mildly interested watcher gives up. The true devotee possesses an enthusiasm that burns so fiercely it carries him over the uneventful between-times when nothing is happening.

SEPTEMBER • 17

BEAUTY IN DISGUISE During these last of the summer days, I see caterpillars spinning their cocoons among the trees and bushes and vines of the Insect Garden. I stopped for a time this morning to examine the brown, paper-like shell of a cecropia cocoon anchored to the branch of a wild cherry beside the swamp path. Within, behind the curtain of this shell, there will take place that incredible transformation in which ugliness is left behind and the repulsive larva, tissue by tissue, is transformed into the superlative beauty of the adult moth. Beauty will come from beauty in disguise.

When we draw back in revulsion at the sight of a caterpillar, we are doing, apparently, what nature and the caterpillar want us to do. A caterpillar is one of the most defenseless creatures in the world. It depends upon its ugliness to protect it. Its hideous repulsiveness is the armor of its defense. What the caterpillar needs most of all is to be avoided and let alone. Because most humans shudder and shun it, it has a better chance of survival.

The one purpose in life for these unshapely children of moths and butterflies is to eat and grow. A caterpillar is storing up the tissues which later will be transformed, within the chrysalis or cocoon, into the winged body of the adult insect. Moth and butterfly do all their growing in the larval stage, and most of their eating, too. Once the winged insect appears, it grows no larger. And it eats, in general, only enough to supply the energy consumed in flight. In its adult form, it is a dainty feeder, a drinker of nectar. It is during its larval stage that excess food and new tissues are in demand.

A few caterpillars have sharp, spiny hairs that sting like nettles. A few others have bright spots that give them the appearance of snake's heads or other fearsome objects. But, in the main, it is upon its own ugliness that the caterpillar depends for its protection.

In my well-thumbed copy of Frank E. Lutz's *Field Book of Insects*, I sometimes turn to these words quoted in his introduction to the Lepidoptera: "And what's a butterfly? At best, he's but a caterpillar, drest." The grace and elegance of every adult butterfly and moth, as it flits about on wings radiant with color, are counterbalanced by the grossness and ugliness of the immature insect. The disguise of this beauty in disguise is so effective that the caterpillar is known the world around as a synonym for the repulsive. It is a repulsiveness, however, that often saves the larva's life.

SEPTEMBER · 18

SPIDER WEBS Heavy dew this morning and every spider web in the garden is strung with pearls of moisture. There are tiny webs among the twigs of dead branches on the apple trees, the orb webs of the golden garden spiders among the higher weeds and grass tangles—webs wherever I look, all shining things of silver beauty. The difference between utility and utility plus beauty is the difference between telephone wires and the spider's web.

SEPTEMBER · 19

THE PERFECT MENTAL IMAGE I have been remembering a little incident that occurred one summer day, years ago. I was on the Massachusetts coast, watching shorebirds with Ludlow Griscom, whose skill at rapid field identification is legendary. He called off greater and lesser yellowlegs without an instant's delay. "How," I wanted to know, "do you tell them apart at a glance?" His reply was: "It is largely a matter of having a perfect mental image of each bird." To tell the truth, at the moment, that sounded like the kind of answer given to child to keep him quiet. But upon reflection, it seems profoundly true. Most of our mental pictures are blurred and fuzzy. We see approximations instead of clear images. We fail to appreciate differences because we fail, with mind and eye, to see what we are looking at clearly in the first place.

SEPTEMBER • 20

NIGHT DOESN'T FALL These latter days of summer mark, with their visible changes, a swifter descent in the long decline of the season. Smoothly, like water picking up speed as it nears the millrace, the days reveal the impetus of approaching autumn. In the noon sunshine, the insects sing with summer cheer. But, even at midday, there is a slight chill in the shadows. I can sense a change, a shift in balance, a premonition of swifter change to come.

We see the end of this day far out on the sea meadows. The light ebbs away in a glory of pastel colors. The change from day to dusk to dark comes in an almost imperceptible transition. But before we reach home, nightfall is complete.

We call it nightfall. But night doesn't fall.

It doesn't descend from the sky so much as it rises from the earth. Similarly, we speak of falling dew. Neither night nor the dew descends out of the sky. The dew condenses from moisture near the ground. And darkness begins at the earth and ascends. The shadows increase in number and density. Night is the sum total of the shadows. It is one consuming shadow in the end—the shadow of our earth itself.

Anyone who has ridden in an airplane at sunset or as the twilight deepens, knows that night rises from the earth instead of descending from the sky. He can see the gathering of night on the ground below him. Sunlight still bathes his plane in brilliance when the hills and fields, the woods and lakes and streams below have faded into the deepening gloom. Only after night has established itself on the surface of the earth does it ascend into the upper sky.

The pace at which the dusk advances depends upon where you see it on the ball of the earth. Near the equator, the transition between day and night is relatively swift. The twilights are short. But farther from the equator, to north and south, the evenings grow longer. Night comes gradually; sunset lingers; and the dusk of summer days is a slow, barely discernible transition. When does twilight end? Meteorologists have provided an exact answer. Night has come and twilight is over when the sun is eighteen degrees below the horizon.

SEPTEMBER · 21

THE FLOCKING BIRDS Ten thousand tree swallows, at least, swirl and sweep and drift like clouds of smoke as we follow the Jones Beach parkway to Captree today. Beyond the Tobay sanctuary, two hundred or more black-crowned night herons mill about over the tree tops. Everywhere the flocks are growing. Everywhere the birds are slipping away by day and by night. During these weeks, species after species, flock after flock, fare forth in the great adventure of fall.

The swallow flocks are loosely knit. They have none of the compact, disciplined movement of the shorebirds. The swallows fly wildly, unpredictably, feeding as they go. One flock sweeps low around us. For a moment, we are engulfed by dodging, skimming swallows. Electric in the air is the excitement of the migrants. We watch them, birds that will never know a winter, moving away down the island, heading for the Jersey coast and the long Atlantic flyway to the south. We watch them with emotions tinged with envy. How free they seem, how fortunate! Part of the fascination that surrounds the southbound flocks is the ever-recurring illusion that the traveler is leaving his troubles behind.

SEPTEMBER · 22

THE ANT'S CAMELS About 2 P.M., in warm sunshine after a light frost in the night, I find close-packed clusters of the brownish tree hopper, *Entylia sinuata*, beneath several sunflower leaves. They are huddled along the midrib like the huts of a village on the banks of a jungle stream. Each is less than a quarter of an inch long. Beneath my pocket magnifying glass, one of the adults expands into a fantastic little insect camel with twin humps rising steeply above its back. I touch it with the tip of my pencil. It shoots into the air with an explosive snap, unfurls its little wings, and sails away.

Among the clusters, there are numerous immature nymphs, even more oddly formed than the adults. Ridges and spines and knobs adorn their black-and-green bodies. They resemble some creation of the glass-blower's art. All have hatched from eggs laid in the plant tissues along the rib of the leaf by females that have remained close by. These gregarious tree hoppers seem almost to have a family life. Old and young live in close association. All are nourished by the sap of the leaf on which they dwell. As a by-product of their sap-drinking, the nymphs, like the aphides, give off sweet honeydew. And, like the aphides in another respect, the tree hoppers are attended by ants. If, as Linnaeus declared, the plant lice are the milk cows of the ants, the tree hoppers might well be called their camels.

On guard beside one little colony of these insect camels, I see one of the black, gallery-making carpenter ants, *Camponotus pennsylvanicus*. The cluster of tree hoppers contains half a dozen adults and nearly twice as many nymphs. I sway the leaf violently up and down. The ant guard pays no attention. This is something normal; the wind often sways the leaf. But when I snap the leaf smartly with a forefinger, the ant rushes into action. This is something abnormal. Alert for enemies, it circles and zigzags over the leaf, at intervals hurrying back to the placid, sap-drinking tree hoppers to go over them hastily with its antennae like a shepherd counting his sheep to be sure none is lost. Similarly, every time I photograph the cluster on the leaf, the sudden, blinding stab of light from the photoflash bulb sends the guardian ant racing about in search of an enemy.

During the hours I watch events on this one sunflower leaf, no real foe appears to molest the membracids. Part of the time, one ant is on guard; part of the time, two. Once, when both the black ants are temporarily away, tiny brown ants swarm over the leaf and feast on the honeydew. But, when I look again, only a few minutes later, a carpenter ant is back on guard and the small ants have disappeared. I see one of the little ants farther down the sunflower stalk. I pick it up and drop it in the midst of the tree hoppers, where the coveted honeydew is thickest. But it stops not for a single mouthful; it scurries away at top speed. I recall the frenzied flight of the carpenter ant I dropped on the aphid-infested milkweed guarded by large reddish ants. The tree hoppers of my sunflower leaf and the aphides of the milkweed apparently are recognized as the property of the ants in possession. They are the herds of the possessors—part of the food supply of the colony.

Bending close, I watch the ants on guard milk the nymphs, stroking their backs, as ants stroke the backs of aphides, to induce them to give off droplets of the sweet honeydew. Research has shown that plant lice milked in this way suck larger quantities of sap than those not attended by the ants. In the course of a single day, one aphid was observed to give off 48 drops of honeydew. On occasions, a plant louse will produce drops of sweet fluid for several different ants in relatively quick succession. Normally, plant lice shoot the sticky drops a little distance from their bodies; but, when they are stroked by ants, they give off the honeydew slowly. In fact, some aphides that are known to have been associated with ants for millions of years, have a little circlet of stiff bristles that conveniently hold the honeydew drop until the ant can take it.

Watching events on the sunflower leaf as the afternoon advances, I see a similar occurrence take place over and over. As soon as a second ant joins the one on guard, the two insects place their mouths together and some of the collected honeydew is passed to the newcomer. I begin to suspect that most of the work of guarding and milking the tree hoppers is being done by the same ant. Am I right? The only way to tell is to mark the insect. I touch it with the tip of a fine brush dipped in white enamel. That is sufficient to leave one side of its inky abdomen gleaming white. I can recognize this ant as far as I can see it. But I do not have a chance to recognize it for long. Startled by the touch of the brush, it loses its hold and tumbles into the grass tangle beneath the sunflower. All the rest of this day, the last full day of summer, the tree hoppers go unattended. Will I find the black-and-white ant on guard tomorrow?

SEPTEMBER · 23

DEWDROP PRISM A single drop of water, at the tip of a twig, reflects the sunlight as I move up the hillside today. I see the drop, acting as a prism in a manner I have never noticed before, shine a brilliant blue. A step farther on, I see the same drop shine with a greenish tint. Another step and it is a brilliant, glittering yellow. Over and over again, I retrace my steps to see this sequence of shining colors. Only along the line of one particular angle can the prismatic effect be observed. It is as colorful as a rainbow. This one drop of dew, seen from different angles, takes on the appearance of a whole series of vivid and striking gems.

THE BLACK-AND-WHITE ANT Under the sunflower leaf, the insect camels are clustered along the midrib as they were yesterday. And standing guard over them, on this day when autumn begins, is the same ant, now resplendent in black and white.

During the day, I look at the leaf repeatedly. The guardian ant is there most of the time. Occasionally, it will disappear for half an hour or longer. But, in the main, tending this band of insect camels seems to be its special job. It is there when I go home for supper and it is there when I look again at bedtime. Does it stand guard through the night? To find out, I get up after midnight and come back with a flashlight. I run its beam over the leaf. All the tree hoppers are motionless, apparently sound asleep. And there beside them, its white patch spotlighted by the beam, slumbers their steadfast guard.

SEPTEMBER · 24

FUNGUS TIME Now is a time of fungus at the Insect Garden. I see new puffballs appearing overnight. Toadstools have pushed up along the swamp path during the latter days of summer. And, this morning,

I pause beside a wild cherry tree to admire the beauty of shelf fungus that has sprung, in a series of descending steps or cascades, from a portion of its trunk. The tree is dying, its top broken away, its midlimbs leafless. The fungus has grown since last I passed the tree. There are ten of the fungi, forming five overlapping tiers. Each of the ten is shaped like a flat, whitish biscuit with brownish markings radiating out over the top, regularly arranged like overlapping feathers or scales on a fish. Here is beauty from decay, a frail and insubstantial form of life, a kind of botanical ectoplasm appearing from the bark of the dying wild cherry tree.

CRUMBS A migrant brown thrasher has spent the day in our backyard. I see it, about four P.M., under the cedar tree by the kitchen window. A bluejay has carried a crust of bread to a limb of the cedar, about six feet from the ground. Holding the bread on the limb with one foot, it pecks at it. Fragments fall to the ground. There, the brown thrasher, hopping about, dines on these crumbs from the bluejay's table.

SEPTEMBER · 25

A MANTIS GROWS UP Among the rose bushes this evening, I come upon a praying mantis just finished with its final molt. Its newly formed wings are soft and filmy. It clings motionless while its chitin shell hardens. Beside it, dangling head downward and secured to the rose bush by one ghostly shell of a foot, is the skin from which it has emerged. With this final molt, the mantis grows up. It attains its wings and its adult form and the last of its external skeletons. It will change no more.

THE MARKED ANT Today, as yesterday and the day before, the marked carpenter ant is guarding and milking the cluster of *Entylia sinuata* tree hoppers beneath the sunflower leaf. These sap-drinkers and honeydew-producers may represent one of the outpost herds of the ant colony that is mining within the trunk of the old Lincoln Tree. I see the black faces of these carpenter ants appear at openings in the trunk where a lower limb was cut away. Their jaws are gripping bits of excavated wood. Each ant leans far out, opens its jaws and drops the wood. Thus, grain by grain, the refuse from the invisible sawmill of the ants collects on the ground below. It spreads out, for a foot or more, over webs that spiders have spun in the grass. These silken traps have captured not insects but sawdust. During the mid-afternoon warmth of these late September days, the insect sawmill within the tree is running at top speed. This is the largest carpenter ant colony in the Insect Garden. How far do its workers range in their search for food? The sunflower leaf is about forty feet from the Lincoln Tree.

SEPTEMBER · 26

YE SUNNY GLADES! I have been sitting on the grass in a sunny little glade among wild cherry trees. And I have been staring at the most profoundly baffling thing in nature. It opens a dark door of thought and brings to the mind a chill, even in this sunny glade.

Hovering, swinging, descending, going from leaf to leaf, an ichneumon fly searches for its prey. All summer, I have seen these relatives of the wasps on such a hunt. They lay their eggs in the bodies of caterpillars where their children slowly devour their hosts. Endowed with fiendish cunning, some of these consuming worms wait until the very last before they touch the vital organs, thus letting their doomed host linger on as a source of food for a few days longer. Contemplating this instinctive cunning, this unconscious cruelty—which was devised by neither wasp nor worm—the thoughtful person finds a somber shadow drawing across his mind.

Those who walk abroad and see only beauty in nature, only happiness in the singing of birds, only beneficent kindness running through all the natural world—they have never really seen this world at all. Ye sunny glades—look again! Look more deeply! See more truly! The unpleasant, too, is part of the world the honest naturalist must see. The padded paw is for bloodthirsty killing and the poison fang for sudden death. They cannot be ignored in his philosophy. They are as much an inherent part of nature as the sunshine and the flowers.

His goal must be the goal of Thoreau, to be able to give a *true* account of the world as it is. But, always, behind the world as it is lies the shrouded riddle of the origin of cruelties that are instinctive.

In the last pages of his classic *Natural History of Aquatic Insects*, the British entomologist, L. C. Miall, sets down reflections on this enigma that are well worth repeating here. "What may be the solution of the mystery," he writes, "and how so much benevolent foresight can be reconciled with so much cruelty, it is not for the naturalist to explain, though the mere naturalist finds it hard to shake off these thoughts when they once have come up in his mind.

"The beginnings of mercy and unselfishness appear in the behavior of wild animals to their young. Social animals are called upon for a more habitual exercise of the same attributes, and it is perhaps the social state which has chiefly moderated the selfishness of man; it is the social instinct which leads him to pity even the humblest victim of the struggle for existence.

"When we have to tell what we have seen and found, it is our business to give a true account, disguising nothing, and keeping nothing back. But let us be careful not to speak as if our little plummets had sounded the depths of the universe. Those who have surpassed their fellows in the improvement of natural knowledge, have always been the first to admit that what they have come to know is lost in the infinitude of the unknown."

SEPTEMBER · 27

TAUT SILK This morning I come upon a thread of spider silk, straight and taut, stretching between two adjacent apple trees at the Insect Garden. One end anchored almost exactly opposite the other end, the fine, glistening line bridges a gap of eight feet or more about seven feet above the tangled grass. How has the little spider stretched the thread across the open air? Only one hypothesis—and it is the true explanation—solves the riddle. Sitting at the very twig tip of a branch, the spider lets the silken thread play out in the breeze. Even the slightest movement of the air is sufficient to carry the spider's silken cable. When the end of the drifting silk becomes entangled with the leaves and twigs of the opposite tree, the spider pulls the line taut, makes it secure, and thus is equipped with a roadway in the air from tree to tree.

SEPTEMBER • 28

REDWINGS BEFORE A STORM As I look upward beside one of the apple trees this afternoon, I see the little redbodied dragonflies alighting on the leaves and taking off again at various levels all up the outer, sunlit edges of the tree. The small dragonflies seem more numerous than ever. But the large species are few now. These strong flyers, like the Monarch butterflies, often migrate down the coast in the fall. Two days ago, among the low dunes of the ocean front at Jones Beach, Nellie witnessed a spectacular movement of such dragonflies. They first appeared in large numbers about noon. All were heading in the same direction, along the aerial trail followed by the Monarchs. Some of the large dragonflies were more than 100 feet in the air. Their numbers were so great that, at times, they paraded past one low dune at the rate of almost 200 a minute.

A little after five this afternoon, long gusts sweep the hot, stagnant air from the hillside. They bring cooler weather, minute by minute, to the garden. Dry leaves bump and patter as they descend among the branches and are snared by the grass clumps below. Curious slate-blue cloud formations curl downward like aerial spray above the eastern horizon. They unfold rapidly like cumuli forming in reverse. The whole cope of the sky is now darkened. Around the horizon runs a narrow band of light, giving the sky the appearance of the top of a pie cut away and separated from the rest of the crust. Blackbirds, that have been away feeding in the fields since sunrise, return in a chattering cloud just ahead of the storm. They swirl downward into the protection of the phragmites. For a moment, the calling of the multitude is stilled. And in the stillness, a single redwing gives the long, rolling "Okaleee!" that carries me back to earliest spring. Then the first big drops splash down and the smell of the hot dust is in the air.

SEPTEMBER • 29

THE STRENGTH OF THE CATTAIL On windy days, for years, I have watched gusts sweep through the cattails of Milburn Swamp. The slender green leaves twist, flutter, bend and gyrate. But they rarely break. And when the wind dies down, they return to their former position. Although such leaves sometimes rise 100 times their width and more than 500 times their maximum thickness, although they are featherlight and partially filled with air, they are able to withstand innumerable stresses and tensions. They represent one of nature's marvels of engineering.

Today, I brought home a dozen or so of the cattail leaves. For an hour, now, I have been busily dissecting them with a razor blade. The secret of their lightness and strength lies in their internal construction. In the swamp, there are two species of cattails—the broad-leaved *Typha latifolia* and the narrow-leaved *Typha angustifolia.* In both species, the internal bracing of the leaves takes the same form. Millions of years before the Wright brothers flew at Kitty Hawk, nature had devised, in the leaf of the cattail, the maximum lightness-and-strength construction of the airplane wing.

Running the full length of the cattail leaf, from base to tip, are veins that are comparable to the longitudinal spars of the aircraft wing. Slice a razor blade across a leaf, cutting it cleanly in two, and the exposed cross section looks exactly like the rib of an early airplane. The upper surface is more curved than the lower and the two surfaces are separated by vertical bracing members. Moreover, when I take cross sections at various points along a leaf five-eighths of an inch wide and seventy-two inches long, I find that, also like the airplane wing, the camber and thickness decrease from base to tip.

For something like 20,000,000 years, the cattail leaf has been making use of this construction that provides it with the greatest strength for the least weight. Paradoxically, the strength of the cattail leaf occasionally works to the disadvantage of the plant. Cattails are pioneer land builders. Their interlacing roots push out into deeper water along the edges of shallow ponds. The advance clumps are insecurely an-

chored and sometimes meet disaster when they are capsized b
wind. The internal bracing of the lightweight leaves is suffici
strong to withstand the gusts. They bend but do not break. In this straining tug of war, it is the massed roots below rather than the leaves above that first give way.

SEPTEMBER • 30

NEWS OF THE NIGHT On this last night of September, Nellie and I wander among the shadowed trees of the Insect Garden, among the morning glories and the mallows and the sunflower stalks. It is before the rising of the moon. Fireflies blink and flash among the dark masses of the swampside bushes. And, from the darkness all around us, rises the music of the night insects. Cricket and conehead and katydid are scraping wing on wing. The sounds that merge and intermingle range from a faint ticking to the ear-piercing, bandsaw song of the conehead. Over our heads, the night is filled with mellow, pulsing waves of sound, the rhythmical music of the snowy tree crickets in the tree tops.

We swing the twin fingers of our flashlight beams this way and that. They spotlight a snowy tree cricket on an apple leaf, a gray sowbug creeping along a little valley in the bark, the black face of a carpenter ant appearing with still another grain of wood in its jaws, a chocolate-colored moth hovering over the butterfly bush with eyes glowing like rubies in the light. Lurking in the buddleia blooms, revealed by our probing beams, are crab spiders, ghostly white, and ambush bugs, immobile and camouflaged. We are wandering at leisure, seeking news of the night.

All across the hillside weed lot, south of the orchard, goldenrod blooms and black field crickets leap away before our feet to bump and thread their way through grass clumps that are suddenly bright in the glare of our flashlights. September—the month of crickets and goldenrod—is almost over.

CHAPTER TEN

October

OCTOBER • 1

GILBERT WHITE'S TORTOISE AND MY CAT Now the mornings break with increasing chill. The coolness of the night carries farther and farther over into the day. At the same time I have noticed an interesting change in the habits of our pet cat, Silver. Throughout the summer, I would find him curled up for an after-breakfast nap flat on the driveway beside the rose garden. In recent days, he has shifted his position a dozen feet or so. Now he sleeps each morning in a little nest made where the rock garden drops down to the driveway on a steep slant. This tilts him toward the east and gives him the benefit of the more direct rays of the sun. I recall now that last autumn he took to sleeping in the same place, although the significance then escaped me.

Gilbert White, a century and a half ago, noted a similar autumnal change in habits by the old Sussex tortoise which he kept in his garden at Selborne, England. "As he avoids heat in the summer," White notes in *The Natural History of Selborne*, "so, in the decline of the year, he improves the faint autumnal beams, by getting within the reflection of a fruit-wall: and, though he never read that planes inclined to the horizon receive a greater share of warmth, he inclines his shell, by tilting it against the wall, to collect and admit every feeble ray."

OCTOBER • 2

ANTS AS WEATHER PROPHETS About five this afternoon, ants swarmed from the ground near the kitchen door. The winged males and queens ran over the grass while the smaller, wingless workers milled about them. A late autumn dispersal flight was commencing. Among the insects, the ants are the best weather prophets I know; every dispersal flight I have ever seen has been followed by at least twenty-four hours of clear weather. But now the sky was darkening. A storm seemed imminent. The insect weather prophets apparently had made a mistake. The ants seemed to come to the same conclusion. For fifteen minutes after we saw them pouring from the ground, we saw them pouring back into the ground. I wondered why this late dispersal flight reached even the door of the nest when conditions were unfavorable. This is the first time I have known winged ants to appear, change their plans, and return into the ground again. They were second-guess weather prophets. And their second guess was correct. Before morning, it was raining.

OCTOBER • 3

THE APPRECIATOR To the appreciator should go the opportunities for appreciation.

OCTOBER • 4

BIRDS THAT SPEAK OUR LANGUAGE In an open field close to the sea meadows, this morning a quail repeated its cheerily whistled "Bob White! Bob White!" And, in trees beyond, two bluejays were screeching

"Thief! Thief! Thief!" as they pursued each other among the branches. A number of bird voices seem to shape themselves into words we know and, as I walked on, I began running over in my mind a list of the birds that talk our language.

Everyone who has walked in the New England woods on a summer day knows that ringing call that echoes among the trees, the "Teacher! Teacher! Teacher!" of the ovenbird. For so small a bird, a member of the warbler family, its voice is surprisingly large and carries surprisingly far. Other warblers have seemed to many to speak our language. The chestnut-sided warbler is supposed to say: "I wish, I wish, I wish to see Miss Beecher!" and the black-throated green is thought to sing of "Trees! Trees! Murmuring trees!"

To Henry Thoreau, the song sparrow shouted: "Maids! Maids! Maids! Put on your teakettle-ettle-ettle!" In the South, the barred owl inquires: "Who? Who? Who cooks for you-all?" The tattler of the shore, the greater yellowlegs, takes wing with a rapidly repeated "Dear! Dear! Dear!" The olive-sided flycatcher, on its high perch, suddenly calls: "Whoops! Three beers!" And, at sunset, the evening robin sings of "Julia Tea-Leaf! Tea-Leaf! Tea-Leaf! Julia Tea-Leaf!"

Of course, what a bird says depends upon who hears it and the conditions under which he hears it. I remember chuckling over one ornithologist's rendition into words of the call of the alder flycatcher among the oak openings of northwestern Ohio. These sandy barrens, under the summer sun, become parched and torrid. Wandering in this thirsty land, the bird watcher noted that all day long the little alder flycatcher called: "Whiskey! Whiskey!" or "Wish beer; more beer!"

Oftentimes, to different people, the same bird says different things. That incomparably beautiful minor melody of the white-throated sparrow, for instance, sounds to some like: "Sweet Canada! Canada! Canada!" To others it suggests: "Old Sam Peabody, Peabody, Peabody!" Or "Sow wheat, Peverly, Peverly, Peverly!" Or "All day whittlin', whittlin', whittlin'!" Or "Oh, long ago, long ago, long ago!" To a friend of mine, the song of the whippoorwill always sounds like: "Purple rib! Purple rib! Purple rib!"

Like the whippoorwill, a number of birds get their common names from the calls they give. The chickadee is one. Others are the phoebe, the veery, the bobolink and the bird that started me along this line of recollection, the Bob White. Such birds not only speak our language; they make it. Their songs have added new words to the dictionary.

OCTOBER • 5

THE FINAL ANSWER—NOT YET! A couple of months ago, I set down the final answer—dragonflies *do* fly backward. That is what I said. But now I find the final answer is not yet! I have been talking to a famous scientist who has spent a good many years of his life studying dragonflies. He is still uncertain whether they actually *fly* backward. He has seen them move to the rear. He has seen them move backward against the breeze. But always, it seemed to him, they moved back on a downward slant as though they were gliding or coasting to the rear rather than propelling themselves in that direction. Am I sure the insects I watched moved straight back without losing altitude? I am not. How hard it is to see with exactness even what lies before our eyes! So when another summer comes, I will be watching for a dragonfly that moves backward against the breeze without descending!

BEHIND THE WALL Night after night, the small birds have been slipping away. The sounds we heard in the dark above the Maine lake have been repeated in the night sky here. From day to day, I realize with a start that summer friends I have known among the garden trees and along the swamp walk have disappeared. One-Leg is gone. Will we ever see him again? This sudden disappearance of the flocks heightens our interest in bird migration. A curtain falls and we, who stay behind, will never know what takes place beyond it. We will never know what adventures befall, what fate awaits, these friends of summer months. We journey with them in imagination only. And we know that quickened and excited curiosity which led Victor Hugo to declare that the most interesting thing in the world is something taking place behind a wall.

OCTOBER · 6

AUTUMN HAZE To the Insect Garden under a brilliant sun and a cloudless sky. Summer seems back again. But the air is filled with autumn haze. Grass fires and burning leaves and the dust of drier autumn all reduce the transparency of the atmosphere. They also play their part in coloring the sunsets and tinting the harvest moon.

Under the old apple trees, fallen fruit sprinkles the hillside and partly dried leaves slip-slap together in the breeze like lapping water. Ahead of me, red-bodied dragonflies whirl aloft from the yellowed grass tangles. The scent of autumn, the smell of wet leaves and the faint, vinegar odor of the fallen fruit, are in the air around me. Now, all over the garden, the larvae have ceased their feeding; the leaves are safe at last.

Beyond the viburnum clumps, I hear redwings, taking off and landing, their monosyllabic "checks" resembling the clacking of tongues. One of the carrion beetles, *Silpha americana*, bright yellow and brown, is clinging to a wild cherry leaf three feet or more above the ground. Its antennae are stretched out to the limit. This is its elevated post, not a lookout post or a listening post, but a smelling post. Perhaps it is trying to catch the scent of the dead bluejay I come upon farther down the swamp trail. Yellowjackets have already found it. These ravenous wasps are everywhere searching for food these autumn days. They are around the bait cans of fishermen and they follow a fruit and vegetable truck that peddles produce down our street.

For a long time, with the sun warm on my back, I stand by the stagnant water of the slow swamp stream. The algae, on this October afternoon, form irregular patches on the brown water. They suggest continents seen from the stratosphere. Between them are the straits and gulfs and inlets of the miniature seas that wash their shores. As I watch, a tidal wave takes form and sweeps toward opposite coastlines. Then a continent trembles in the grip of an earthquake. Up through its center pops the head of a painted turtle. It is the sea disturber, the continent shaker. Here, it occurs to me, is the reenactment of an ancient allegory. In the age of mythology, people believed that Atlas supported the world and that a turtle supported him. Earthquakes were accepted by the myth makers as merely movements of the turtle. Here, before my eyes, the turtle moved and a continent shook.

A little later, other earthquakes were produced when two painted turtles, pursuing and pursued, wound their way—with many a pause and rise to the surface—among the scum masses on the water. Their shells were bright and their colors clear. In their wandering course, they came close to the spot where I was standing without seeing me. I was between them and the sun and they were blinded. A few minutes later, they approached again, this time from the side. The instant they saw me, their swift strokes drove them down out of sight among filmy castles of green underwater algae. What world could I imagine more fantastic than that in which the painted turtle lives?

OCTOBER • 7

GUNNERS Hell hath no fury like the self-righteous indignation of the gunner who sees the thing he wants to kill killed by a predator.

OCTOBER • 8

CENTIPEDE LEGS With all its feet scrambling and slipping on the smooth enamel, a common house centipede, *Scutigera forceps*, was struggling to escape from the kitchen sink when I came downstairs this morning. I tried to help but the centipede would have been better off if I had stayed in bed. I made a dab at it, trying to pick it up by one leg. The leg came off between my thumb and forefinger. I tried again and again, each time separating the creature from another of its expendable legs. A centipede does not have 100 legs, only about 30. If I had helped it much longer it would have been a biped. As it was, it finally made good its escape, running off without a limp on a score or more of agile limbs. It left me staring in amazement at the legs between my thumb and forefinger. These fragments of an animal seemed individually endowed with life. They persisted in the struggle to escape. Each leg continued to wriggle and squirm for five or ten seconds after it was separated from the living body of the centipede.

OCTOBER · 9

A MANTIS CATCHES A SHREW At the rear of the yard this afternoon, I noticed a mantis on the ground feeding on something dark which it gripped with its spined forelegs. As I came close, I discovered its victim was a short-tailed shrew. It was four inches long, from tip of nose to tip of tail, or three inches long, not counting the tail. The shrew was dead but still warm. The praying mantis had begun to nibble at the back of its neck, the same spot where it commences its feasts on captured grasshoppers and butterflies. At first I thought perhaps the mantis found a shrew dropped by Silver but I have since concluded that it probably imprisoned the little mammal within the toothed trap of its forelegs and that the high-strung creature had died of fright or excitement in the struggle. Certainly it had been alive only a few moments before I discovered it.

This is the only instance I am aware of in which a mantis has captured a shrew. But *The Journal of the New York Entomological Society*, a few years ago, recorded a case in which a mantis caught and killed a meadow mouse. *The Auk*, the official publication of the American Ornithologists' Union, in the July 1949 issue, reported from two widely separated locations—Texas and Pennsylvania—instances of praying mantes catching hummingbirds. A few years ago, an Associated Press dispatch told of a battle between a mantis and a bat on a ledge of a downtown office building in Columbus, Ohio. The bat lost its life. In a terrarium, in a New York City public school, a mantis once caught and killed a small DeKay garter snake, consuming nearly half of it. As autumn advances each year, and the number of insects decreases, the carnivorous mantes become more ravenous. On several occasions, late in fall, I have come upon a praying mantis consuming one of its own legs—an animal devouring itself to allay the pangs of its terrible hunger.

While I am setting down bizarre items on the unpredictable menu of the mantis, I should note a letter recently received from Burma. A missionary there, who has read some of my books, tells of feeding a mantis drops of milk from the end of a matchstick. It consumed drop after drop, drinking at least two teaspoonfuls before it was satisfied.

OCTOBER · 10

NIGHT OF THE FIRST FROST Stars glittered in a clear sky when I went to bed last night and the mercury was falling. This morning, all the grass tangles are silvered with rime, the first heavy frost of the year. I walk across the silent fields. How many small summer singers have been silenced forever! The weed lot is like a deserted battlefield, a Waterloo, where tiny flames of life were met and vanquished by the cold. The insects not killed by the drop of the mercury are chilled into silence and immobility. They will sing again in the warmth of midday but each succeeding frost will thin the ranks of the insect musicians.

And everywhere, out on the sea meadows, by the bay, in the swamp, by Milburn Pond, the results of the first frost will be apparent. It is now that the bulk of the redwings leave. Blackbirds begin to trickle in from the north as early as late July. The swamp is a way station for the northern flocks as they move south. As late as November, migrant redwings are still on the move and, near the coast, a few stay all winter—invariably males. But it is at the time of the first heavy frost in early October each year, that the flocks of the redwings reach their peak abundance.

OCTOBER · 11

VALUABLE VERMIN We go to the bay meadows to see the scarlet of the samphire. The green of the fleshy salicornia has been transformed, all across the flats, into a resplendently beautiful red. Scattered among the wiry salt meadow grass, the glasswort rises like little branching corals, brilliantly tinted. While we walk over the moor, delighting in this new color and beauty that the frost has brought, two marsh hawks drift back and forth, tacking this way and that in their search for meadow mice. Three or four times, late-remaining sparrows fly up almost beneath them. The hawks pay no attention to them. Their interest is in rodents.

In the days when bounties were paid for killing hawks, professional gunners used various baits to attract the winged hunters within range. They found that they got far more hawks when they used mice and rats than when they employed small birds. The benefit of hawks far outweighs the harm they do. They are an essential part of nature's balanced whole. And so are many other creatures lumped together under the odious heading of vermin.

In England, where generations of gamekeepers on the big estates have spent their lives slaughtering "vermin" in the interest of larger yields of pheasants and grouse, it has now been discovered that a natural balance of animal life produces even better results. In *The Countryman*, for the spring of 1953, I have come across the following interesting report on the value of vermin:

"On estates in Hampshire and Norfolk where there are now no keepers and vermin (so-called) have increased, as many pheasants and partridges have been shot since the war as before it, according to Colonel Richard Meinertzhagen. He also told members of the British Ornithologists' Club that he had personal experience of two estates in Ross-shire, totaling 25,000 acres—mostly deer forest and parts of it at over 3000 feet—where the killing of all vermin except rats and lesser black-backed gulls was forbidden in 1920. At that time, the annual yield of game rarely exceeded 600 brace of grouse and about 100 pheasants and partridges; the only predatory birds were a pair of buzzards, two pairs of sparrow hawks and a few hooded crows. By 1932 a pair of golden eagles bred there regularly, as well as four pairs of buzzards, one of peregrines and two of ravens, and many kestrels, merlins, sparrow hawks and magpies. Since stoats and weasels had been unmolested, the farms had been cleared of rats. Yet the two estates were yielding over 1000 brace of grouse, besides large bags of pheasants, partridges and capercaillie. Moreover, there were more snow buntings, dotterel and golden plover than on other estates of the same size, so that the rarer small birds had not suffered either."

OCTOBER · 12

OCTOBER MIST A luminous, misty dawn. And, all down the hillside, the spider webs shining with dew. Like the stars, the webs have been there all the time; moisture is the night that makes them visible. On such October days as this, we look about us as though in some new and magic land. The mystical draws close behind the luminous veil. We see the things about us and sense larger meanings just beyond our grasp. Looking back on such a time, we add—as Thoreau did one autumn day—"And something more I saw which cannot easily be described."

OCTOBER · 13

THE TERRITORY OF THE MANTIS All day long, on summer days, I have watched a praying mantis cling near a buddleia bloom, never varying its position more than a few inches from morning until night. An explorer friend of mine, at the American Museum of Natural History, once told me of a mantis he watched on a bush in the tropics. It came to the same spot day after day. So closely did it resemble a leaf in form and coloration that it virtually disappeared when it settled down among the foliage. Only when it remained still and the leaves around it were stirred by a breeze did it become apparent. But this was at rare intervals for, as soon as a breeze sprang up and the leaves stirred, the insect would begin bending its knees and swaying from side to side, moving like the moving leaves around it.

Remembering the mantis of this tropical bush and of my buddleia bloom, recently, I began to wonder to what extent a mantis has a definite territory of its own. Is the mantis I see today among the rose bushes the same individual I saw there yesterday and the day before and last week? The only way to tell is by marking the insect. So I got out my fine brush and white enamel and placed little distinguishing dots on various parts of the bodies of twelve praying mantes I found in the yard. Six were males, six females. I drew a little map and noted down where each mantis was hunting when it was marked. In the days that have followed, I have been keeping track of the movements of these branded insects.

One has a white spot on the tip of a wing, another a spot on the top of its abdomen, a third on its thorax, a fourth on its right foreleg, and so on. Exactly half of the twelve wandered away or were killed by birds. I never saw them again after the day I marked them. Among the other six, the most consistent was the fifth one I marked. It remained on or around a small yew for more than two weeks. Here I saw it with captured prey. Here I saw it mating. Here I saw it forming the froth oötheca that holds its eggs. Of the other marked insects, one remained for the better part of a week near a clump of pokeberries close to the kitchen window; another spent four days in a high, plumy clump of grass; a third kept reappearing among the rose bushes. The most surprising thing to me was that four out of the six mantes that remained in the yard were males. My impression is that it is the smaller, lighter males that do the most flying and moving about in the fall. I had expected that all of the more sedentary of the marked mantes would be females. Another time this might be so. For this one test is what a careful scientist might mark: "A Note Preliminary to Investigation." Another summer will see more marked mantes in my yard. That is something to look forward to.

OCTOBER · 14

THE HABITS OF TREES Three clumps of arrowwood rise by the path at the foot of the Insect Garden. One is brilliant red; another part red, part green; the third entirely green. On the slope, one of the wild cherry trees turned red long before its fellows. I recall this same tree was first to be clad in autumn foliage a year ago. Within sight of my study window, at home, there are a number of Norway maples. One tree, year after year, stands out because of the brilliance of its autumn leaves. Even trees are creatures of habit. The one that has the most color in its leaves this autumn is likely to have the most color next autumn. It is a part of the characteristics, or "habits," of the tree.

In the sunset, I walk home along the hillside. More and more, as the days of fall advance, the quiet of the garden at sundown will suggest the empty stage and wings of a theater. The lights are dimming; the audience has left; the actors are trooping away.

OCTOBER • 15

THE SILENT BATTLE Again, the dawn mist is smoky gray beneath the trees, shining, silver gray in the open. As I wander through the Insect Garden, I come upon a woolly bear spangled and starred with droplets of moisture. A little wild cherry beside the trail is adorned like a Christmas tree. Its dead branches are spider-trimmed with ropes and webs and dew-decorated spangles. Above me, a cheery "perchicory!" call marks the passage of a goldfinch, invisible in the vapor. As I approach the swamp stream, out of the autumn fog phragmites and cattails suddenly seem to step forward.

For a decade and a half now, I have watched the silent battle between these two. Slowly the phragmites have extended their foothold. These high, plumed reeds have less food value for wildlife than the cattails. But around the world, where they are found near the coasts in temperate zones, they have many uses. The old masters used to shape their drawing pens from the stems of phragmites. Today, in the south of France, along the Mediterranean, the slender reeds are used for thatching houses. In Milburn Swamp, their chief value is as a shelter for blackbirds, grackles and starlings. Each day, at sunset, I see the great starling flocks come in to roost for the night. As I stand in the chill morning vapor, looking into the phragmites—gray and wavering in the gray mist—I begin to imagine what it must be like within that dense stand of canes when the starling hosts pour down from the sky. An adventure in viewpoint takes shape in my mind. Tomorrow, before sunset, I will return and secrete myself in the heart of the phragmites and watch homecoming birds descend around me.

OCTOBER • 16

IN THE PHRAGMITES A little before five P.M., wearing long boots, old army pants, a leather jacket and an ancient felt hat, I push my way into the jungle of the phragmites. The day has been warm for mid-October. But a chill rises early from the swamp. In the heart of the stand, I tramp out a little hollow. Here I await the returning birds and here I set down the following notes:

5:01 P.M. I can see only four or five feet, at most, into the tangle around me. Here and there, by the laws of chance, the stems are arranged into corridors that open for several feet—high, thin corridors barred at the end by a maze of upright stems. The interior of the canebrake is so like pictures of bamboo jungles of the Orient that a tiger or cobra would hardly seem out of place. Over my head, eight or nine feet above the floor of the swamp, wave the plumes of the phragmites. Each stem is hardly thicker than my little finger.

5:05 P.M. Each time a breeze sweeps across the swamp, there is a creaking of stems and a dry rustle of leaves. The fall is a time of growing brittleness.

5:07 P.M. The first bird to arrive is a redwing. I hear its monosyllabic "Check! Check!" as it alights in the willow and then flies down to the far side of the phragmites.

5:09 P.M. Birds are assembling in the swampside maples. The sun is going down and the flocks are coming in. I can hear the mingled calling of many birds.

5:12 P.M. The clamor increases. The trees are thronging with new arrivals. Seeing little, I depend upon my ears for news. Suddenly there is the airy sound of great numbers of birds taking off, coming nearer. A cloud of redwings, grackles and starlings engulfs the phragmites. The flapping of wings is like a wind. Black shapes go hurtling across the tiny cope of my sky where the phragmites part slightly above my head. The reeds seem shaken by great gusts as birds pitch down from the air among them. The calling of redwings, the metallic voices of the grackles, the whistling of starlings blend into a medley that is almost deafening—the confused clamor of the great flock.

5:15 P.M. After three minutes of this bird bedlam, my ears feel weary and battered and deafened. There are alarm notes, quarreling notes, excitement, discontent, gossiping, the hubbub of the hundreds of birds around me.

5:20 P.M. The uproar reaches a crescendo. Then there is, suddenly, silence. A perfect hush falls on the multitude of birds. It is followed by a silken whirring, a mighty fluttering, and the plumes of the phragmites wave in the wing-formed wind as the birds rise in unison into the air. For a moment, all the stems around me rock as in a breeze.

5:22 P.M. Beyond the swamp stream, the birds have alighted in another and larger stand of phragmites where they will spend the night. I hear their confused clamor muted by distance. I also hear the voices of a new concentration of starlings building up in the maples.

5:31 P.M. The chill is growing. Shadows have engulfed the phragmites.

5:42 P.M. Now the second wave breaks over the plumed reeds. The deserted phragmites are again filled with life. The concentration and the vast tumult builds up with each new addition to the flock. A redwing—one of those still to migrate—alights close behind my head. Its surprised "Check!" rings loud in my ears. There are little pauses, from time to time, in the vast tumult of bird voices beating against my ears. And in one of them a single starling, almost overhead, imitates a snatch of the sweet, plaintive song of the white-throat.

5:51 P.M. Again there is the sudden hushing of the din, the sudden mounting "Wooosh!" of wings, the swaying of the plumes above my head. These birds, too, cross the stream to the larger stand of phragmites for the night.

6:07 P.M. The individual reed stems are merging together in the gloom. The chill of the night, the smell of the swamp, the feel of mist is in the air.

6:39 P.M. No more birds have come. Outside I hear the flutter and splash of two ducks alighting on the swamp stream. The thronging birds in the far phragmites have fallen quiet. Now, even the steely hum of a mosquito sounds large in my ears. I flounder out of the phragmites, feeling my way, and ascend the garden slope. Darkness is all around. The homecoming of the birds is over.

OCTOBER · 17

SPIDER SONGS For a long time now I have been listening for spider songs. But I have never heard one. Spiders are considered stealthy and silent creatures. Stealthy they may be; but they are not always silent. A number of species produce sounds audible to the human ear. John Burroughs was once sitting among fallen leaves in the autumn when he was attracted by a low, purring sound. He was amazed to find it was produced by a spider. All spider songs are instrumental. They are produced by spines that rub together. Most spider music is low and difficult to catch, but I have heard of one species that will stand its ground when disturbed and buzz loudly like an angry bee.

OCTOBER · 18

THE SWINGING LEAF For five minutes this morning, I watched a mantis held prisoner by a swinging leaf. The insect clung to a rose bush just outside the kitchen window. Near it, a dry leaf dangled at the end of a thread of spider silk. The breeze swung the leaf back and forth pendulumwise, and the head of the mantis turned from side to side, like the head of a spectator at a tennis match, as it followed its movements. Anything that moves is of interest to the praying mantis. It is through movement that it is attracted to its prey.

A BLACKBIRD TRIES TO CATCH A MANTIS At the edge of the swamp, this afternoon, a praying mantis launched itself from a dry grass clump and fluttered away along the hillside. I watched a red-winged blackbird take after it, dive under it, come up from below and try to pick the slow-flying insect from the air. But the mantis landed among cattails unharmed and uncaptured. Perhaps its bizarre appearance turned the redwing aside at the last moment.

On a number of occasions, I have seen birds bluffed out by these fearless insects. A few days ago, a bluejay alighted near a mantis in our backyard. It hopped toward the reared insect. The mantis gave no ground. It lashed out with its spined forelegs whenever the bird came near. It turned to face the bluejay each time it approached from a different direction. In the end, the jay flew off. For minutes afterward, the aroused mantis remained erect, reared and ready for another attack.

OCTOBER • 19

THE GIFT OF WONDER What a natural wellspring—cooling and refreshing the years—is the gift of wonder! It removes the dryness from life and keeps our days fresh and expanding.

OCTOBER • 20

THE MIGHTY MITES I came upon a beetle this morning, one of the ground beetles, perhaps half an inch long, laboring across the driveway. It carried with it a bizarre, living cargo. Massed on the underside of its body and overflowing onto its back was a solid mass of mites. There seemed to be literally hundreds of them, crowded and clinging together, several deep. The mass was constantly in motion. The miserable beetle, supporting this parasitic host, moved slowly, painfully, with difficulty. Mites are everywhere in nature. They ride on birds and they travel pickaback on snakes and they invade stored foods and they produce minute galls in plants. There is a red mite that lives on dragonflies, a mite that preys on mosquitoes, even an almost-microscopic white mite that is found on *Drosophila* fruit flies. Among the "littler fleas" of Swift's immortal jingle, mites are the *ad infinitum*. The mighty mites!

OCTOBER • 21

SOUND OF THE FALLING LEAVES Chill rain in the night has stripped down the leaves and I walk for the morning papers along a path paved with maple gold. The gray and white cat that, a few months ago, ran to meet me over the green of fallen maple keys, now runs over the yellow of fallen maple leaves. As I come home, I hear the wild clamor of geese in the sky and see the first arrowhead of autumn heading south.

By afternoon, the day has cleared. Sunshine streams through the threadbare clumps of wild cherries at the garden. I walk about, taking stock of the innumerable changes on the hillside. The clump of golden asters is now dry and brown, the milkweeds, stripped of their leaves, are straight spikes thrusting up from the ground and holding the browning seed pods. Seeds are everywhere. I find them between my fingers when I run my hands through the grass tops. Autumn is a time of accounting, summing up, harvest and inventory.

How small are the sounds produced by the death of a leaf! I see the maple leaves and the wild cherry leaves and the apple leaves lose their grip and descend through the air. I catch faint tappings and flutterings as they hit or rub against the twigs. Some leaves swoop and whirl and scud, others descend slowly, directly like the fall of a snowflake. But each breeze that sweeps through the branches overhead today, brings a rain of leaves descending around me. Each hour brings its visible changes to the garden. Change is a measure of time and, in the autumn, time seems speeded up. What was, is not, and never again will be; what is, is change.

OCTOBER • 22

QUICK STARTER An autumn fly has been buzzing about my study this morning, alighting on the curtains, the ceiling, my desk, the bookcases. It is attracted again and again by the warmth and light of my desk lamp. Each time it alights near me, I wave my arm to watch the suddenness of its takeoff. In darting into the air, its movement is literally quicker than the wink of an eye. This was discovered in a government laboratory in Maryland a few years ago, as a by-product of research that involved high-speed photography of bullets striking metal plates. Just as the movie cameras began to whir in one test, a house fly alighted on the metal plate a few inches from the spot where the bullet struck. The impact kicked the plate instantaneously from under the insect's feet. The movie film recorded exactly how long the fly remained suspended in the air before its wings got into action. It was twenty-one microseconds. Thus, in a fiftieth of a second, a house fly can dart from a relaxed position into full flight.

OCTOBER · 23

PEPPERGRASS Among the browns of the garden—the dry sunflowers, the dead mallows, the remains of the once-purple buddleia blooms—among these varied shadings of brown, the peppergrass remains brightly green. This slender relative of the wild mustard has spread, unassisted, over the hillside. In this time of approaching cold and fleeing vegetation, I nibble on the minute leaves. Each fall, at this time, I take special delight in the smarting, pungent, summery taste, the wild seasoning and condiment of the peppergrass.

OCTOBER · 24

A NATIONAL INSECT I am surprised this morning, to see a late Monarch, left behind by the tide of migration and flying south alone, pass over the swamp and hillside. What other insect is so widespread in the United States as the Monarch butterfly? I have seen it in Florida and along the St. Lawrence and in Minnesota and at the mouth of the Mississippi, on the Texas coast and in the mountains of California. It is familiar to almost every part of the country. And its fall migration flight has made it famous throughout the world. It is more beneficial than harmful. It is an insect of the New World. It, among all the insect hosts, might well be chosen as our national insect. Thoreau justly complained that legislatures would appropriate money for the study of only *injurious* insects. The Monarch, beautiful, interesting, and in the main beneficial, deserves the attention of at least one legislature. We have state birds and state trees and state wildflowers. Why not a state insect? I nominate the Monarch!

OCTOBER · 25

SMOKY DUSK To Milburn Pond at sunset. Crimson hips are thick on the wild rose tangles and the muskrats are adding new layers to their houses. I see a bluejay drinking from a little pool of collected water in a knothole. Pale gnats hover around me in the growing chill of the twilight. Continuing on after so many larger and stronger insects have been killed by the frost, they are living examples of the endurance of the frail. The smoky dusk deepens, settling down in varied grays, and I walk home through a nightfall redolent with the nostalgic autumn smell of burning leaves.

OCTOBER · 26

THE ARROW-SHOT DUCK There is a little lake in a small park near here where several white ducks paddle about waiting to be fed by visitors. A watchman at the park recently pointed out one of the largest and healthiest of the ducks and related the astonishing adventure which befell it some time before.

He found it, one morning, the victim of some degenerate with a bow, pierced by three arrows, one through its tail and two in its right thigh. The watchman—who observed that he thought "the duck was a gone goose"—pulled out the arrows. He said they "were meated up and left holes." He expected the bird to die in a few hours. For days it lay with hardly a movement. Then, for about two weeks, it floated in the water, getting about with difficulty. At the end of that time, it began to pick up and made a complete recovery.

Unless a vital organ is struck, birds often appear able to endure amazingly severe wounds. *Life Magazine*, in the spring of 1953, published a photograph of a herring gull that flew about and lived a normal life with an arrow piercing its back and projecting from either side of its body just behind the wings. Some years ago, *The Auk* carried an account of a robin which, through some accident, had been impaled by a stick that pierced its back and protruded from its breast. In spite of this seemingly mortal injury, it lived at least two years afterward.

Not only did it live but, with the stick thrust through its body all the time, it mated, helped raise two broods and apparently migrated.

OCTOBER · 27

A DOG THAT LIKED FLOWERS Florence Page Jaques once told me of a dog she knew that was fascinated by flowers. It was a cocker spaniel belonging to a friend of hers. All summer long it was continually poking its nose into blooms or standing on its hind legs to see bouquets on tables. Whenever it went into a strange room, it headed directly for any flowers there, running around them and stretching out its black nose the better to smell the blooms. So far as I can recall, this is the only dog I ever heard of that revealed any interest in the perfume of flowers.

OCTOBER · 28

THE SEQUENCE OF FALLING LEAVES Along the swamp trail to the garden at midday. Woolly bears are out on the open path. The wind is boisterous but not cold. It is like an overgrown Newfoundland puppy, bowling you over in the exuberance of its spirits. Leaves sail by and silken seeds. The forms of the summer insects—the green gnats and the honey-yellow hoverflies—are now replaced by floating seeds. I turn over a board lying in the grass. A month ago, crickets would have leaped in all directions. Now I see but two or three. Apparently, most of them die afield, overtaken by the autumn cold while they are abroad at dusk. Up the slope of the garden, fallen apples and pears, dull golden-yellow and honey-sweet, lie hidden in the grass—a feast for ants and yellowjackets. There is a scratching in the fallen leaves and I see the red tail of a migrating hermit thrush beneath the wild cherries. The leaves on the branches above it are streaming and fluttering in the wind. Many are being carried away by the gusts. Each autumn there is the same sequence of the falling leaves. The swampside maples are now nearly bare while the weeping willow is clad in almost all of its slender leaves. The boughs of the apple trees are virtually leafless, but the pear trees retain most of their foliage. And, beside the swamp path, the arrowwood that was the last to show autumn color is still clad in the wine-red of its leaves. Some years the sequence begins earlier, some years later. But the order in which the trees of my Insect Garden lose their leaves remains the same.

OCTOBER • 29

NATURAL HISTORY IN A BALLOON The chances are a million to one that if I were asked to list interesting places for the study of natural history I would never think of including a balloon. Yet unusual observations have been made from the wicker baskets of these lighter-than-air craft.

I have been remembering the two U.S. Navy balloonists I once interviewed in connection with a magazine feature. We were in one of the great hangars of the balloon school at Lakehurst, N.J., when our conversation drifted to the reaction of animals to the silent gas bags passing overhead. Almost all animals, I was told, are much more alert to what is going on around them than men are. A balloon will float directly over men at work in a field and oftentimes they never see it until the aeronauts above shout down to them. But the farm animals spot it from afar. Horses sometimes run to the other side of a pasture as the balloon approaches and cows, of a more placid disposition, lift their heads from grazing and watch the gas bag go by.

But the farm dwellers that show the most excitement are the chickens. They begin dashing about the poultry yard and setting up an outcry when the balloon is still in the distance. Even one floating at high altitude catches their eye and causes alarm. The craft apparently is viewed as some kind of mammoth hawk. The navy aeronauts told me they had heard the cackling of alarmed hens more than a thousand feet in the air. The sound that seems to penetrate farthest into the sky, of all animal-produced noises is the yelping of a dog. Alberto Santos-Dumont, the Brazilian pioneer aeronaut, once told of hearing the howling of dogs in an invisible village below him when he was floating high above the clouds in France.

One thing the two balloonists I talked to had been particularly struck by was the actions of carrier pigeons which they released in the air. Such birds were part of the equipment on every free ascent. In case of emergency, they could be released to carry messages. The men had noticed a curious thing about these pigeons. When they were set free so they had to fly *north* to reach home, they attained their goal more quickly and surely than when they had to fly south. Why? The men did not know. It was a natural history riddle they had encountered in a balloon.

OCTOBER · 30

BOOKS A day of cold rain and laboring gusts beneath a sodden sky. I spend the hours indoors among well-loved books ranged shelf on shelf around my study walls. I am in the company of the wise. And I let three of the wise write my entries for this day.

J. Henri Fabre: "Let us dig our furrow in the fields of the commonplace."

Thomas Huxley: "The smallest fact is a window through which the infinite may be seen."

Gilbert White: "All nature is so full that that district produces the greatest variety which is the most examined."

OCTOBER · 31

THE SUNDOWN OF THE YEAR This month comes to a close with a day of lingering sunshine. The rays are genial, the breeze mild, the temperature seventy-five degrees—summer at October's end. I roam the hillside among the goldenrod, now dead and dry and capped with silver. In the garden the dry rustle of leaves, stirred by the breeze, has taken the place of the insect music of only a month ago. Most of the crickets are gone. The clock of their little lives has run down, never to be rewound. At sunset, the breeze dies. All sounds are low or short or subdued. This is the sundown of the day and the month. It is sundown for the year as well.

CHAPTER ELEVEN

November

NOVEMBER · 1

THE ENCHANTED LABYRINTH It occurred to me today that an interest in nature leads you into a kind of enchanted labyrinth. You wander from corridor to corridor; one interest leads to another interest; one discovery to another discovery. It matters little where you begin. You may first fall under the spell of birds or wild flowers or you may become curious about lichens or grasshoppers or trees or rocks or fossils or waterweeds. If you have any inquisitiveness at all, you soon find yourself branching off, wandering enchanted down charming bypaths.

Rowland R. McElvare, a banker friend of mine, some years ago became interested in collecting day-flying *Heliothid* moths. His primary concern was insects. But soon he found he was also studying the plants on which the moths were discovered. He thus developed a liking for botany. Then he began to note the soil in which these plants were most often encountered. In this manner, an understanding of geology began. Like the house that Jack built, his interest in nature continued to expand.

Another friend of mine spent years gathering facts about the early history of the community in which he lived. He spent weekends deciphering worn inscriptions on old gravestones. And, in so doing, he began to notice the moss and lichens. He took up the study of these fascinating forms of plant life on the side. Then his curiosity was aroused by the small beetles and other tiny creatures he found living amid the mosses. In this way, his interest grew and widened.

J. Henri Fabre, world-famous as a student of the insects, began his study of nature by collecting shells when teaching school on the island of Corsica. Later, at Avignon, he searched the fields for toadstools and prepared exquisite drawings of the fungi of Provence, water colors that are still in existence. His fame rests on his study of insects but his interests embraced all nature. Louis Agassiz, noted in far-ranging fields but probably most of all for his work on glaciation, began with the study of fishes. And Charles Darwin entered the enchanted labyrinth of nature study through an early interest in beetles.

We cannot watch sandpipers, following receding waves and stabbing at the wet sand for food, without soon wondering what they are eating. And that leads us into another of nature's innumerable corridors. All living things are linked together in various ways—by predator chains and food chains, by parasitism and symbiosis. Nothing lives to itself alone. Nothing is disassociated from its surroundings. An interest in tadpoles sooner or later leads to an interest in the dytiscus beetle that preys on them. The study of hognosed snakes inevitably leads to the study of toads upon which they feed. Even if your special interest is catbirds, you are led to a consideration of the life of the deer mouse. For these mice often appropriate the nests of catbirds as winter homes. Everywhere we turn in nature, new and interesting corridors appear before us, waiting to be explored. All are interconnecting. They lead us as far as we care to go.

NOVEMBER • 2

SIMILES Green as grass. Quick as a squirrel. Hard as a rock. Slippery as an eel. Busy as a bee. All the truest similes of nature are clichés. Many people have been struck by the same comparison.

NOVEMBER • 3

THE DARKER DAYS Now the hours of daylight are shortening noticeably. Dusk and dark advance minute by minute as these November nights close down. By midafternoon today it is beginning to resemble sundown under a heavily veiled sky. The hundred redwings that apparently are overwintering in the swamp this year, come home from the fields at 3:30 P.M. I see two flocks approach from different directions, flying at the same height. Just before they collide head-on over the swamp, one flock rises, the other descends, and they pass as though they are two single birds meeting in the air.

NOVEMBER • 4

MANY WORLDS The natural world we know is, in reality, a multitude of worlds. As Zangwill, the novelist, once put it: "The scent world of dogs, the eye world of birds, the earth world of worms, the water world of fishes, the flesh world of parasites, intersect one another inextricably and with an infinite interlacing, yet each is a symmetric sphere of being, a rounded whole, and to its denizens the sole and self-sufficient cosmos." We, who live in the world of men, can cross the boundaries of these little worlds around us only in imagination. For we, too, are prisoners within our own individual spheres of being. We look in the mirror and see our faces. They are the only faces we will ever have. We hear our voices. They are the only voices with which we will ever speak. We contemplate our hands. They are the only hands we will ever use. Each of us is, in a manner of speaking, under life sentence, a prisoner within the cell of ourselves. We escape only, as we travel into those lesser worlds around us only, in imagination.

NOVEMBER • 5

INDIAN SUMMER The lowering days have disappeared. Suddenly sunshine pours from a balmy sky. The wind, the raw November wind, has become gentle and bland. Haze encircles the horizon and fills the air with the glowing, magic light of early fall. I go about the hillside with sleeves rolled up. The temperature rises to eighty degrees. Life has returned to the garden. I see honeybees abroad, a red-bodied dragonfly, and gnats drifting in the sunshine. These are halcyon days for the little creatures that have thus far escaped the devastation of the frosts. Milkweed seeds drift through the warm air and late-ballooning spiderlings ride their gossamer threads aloft. All across the weedlot, at sundown, I see their silk entangled with the dry plants, streaming in the breeze. I stand for a long time gazing across the field. The sight before me carries me backward weeks in time into the earliest days of autumn.

NOVEMBER • 6

ADVENTURE LIES UNDER A BOARD Near the top of the garden slope, a decaying plank lies in the grass. It has been there for years. From time to time, I engage in a small adventure, lifting the plank and peering into the bizarre little world that lies beneath. This morning, under the warmth of the Indian summer sun, I view again this diminutive world with a board for a sky. Gray centipedes and varnished brown millipeds flow away among the debris. Pill bugs roll themselves up like armadillos. Sow bugs, flat and elliptical, slide under fragments of moldering wood. It is as though I were rolling back the calendar, stripping away aeons of time, viewing creatures of a lost world. They suggest the decaying remnants of an ancient line, grown small and living among moldering ruins. Although various appendages set them apart, a sow bug, to the casual glance, might well be an ancient trilobite reduced to miniature proportions. In their history, these humble wood lice have seen whole races of men go by. The mammoths of 50,000,000 years ago are extinct. But the sow bug lives on. Survival is a kind of strength. It is an achievement to have endured. The first goal of the living is to live.

More closely related to crabs than to insects, the sow bug, although it lives on land, still breathes through gills. Hence its preference for damp wood and decaying vegetation. It is a scavenger, dining occasionally on a dead insect, more frequently on decomposing plants. When attacked, it can give off an offensive fluid that repels its foes. Because of its extreme flatness, it is able to slip, cockroach-like, into minute cracks. But it is in its manner of reproduction that the sow bug appears most bizarre of all. The female lays eggs. But she does not deposit them in wood or on the ground. She places them in a pouch beneath her body. Here, much in the manner of a kangaroo or sea horse, she carries the baby wood lice until they are ready to appear and make their own way in the world.

NOVEMBER · 7

THE CHILL OF CHANGE Another morning dawns clear and mild. This little time of tranquil skies, these days of Indian summer—here is the annual opiate that blunts our senses, that dulls our minds to the inevitability of the coming cold. Winter seems immeasurably far away. Yet, before evening falls today, there is a change, a chill in the air. So soon, these warm and drifting days are over!

NOVEMBER · 8

VIOLET SEEDS All along the driveway, where blue violets bloom in the spring, I hear a peculiar, fine, snapping sound this morning. At first, I think it must be some small insect among the leaves. But after Nellie and I stand listening intently for some time, we trace it to the little seed pods of the violets. Perhaps the sudden drop in temperature in the night, after days of warmth, or the amount of humidity in the air on this clear, chill morning, has caused the pods to split. Scores burst open with tiny explosions as we listen. Each time, segments of the elongated pod fold out and down to form a three-pointed star, each point packed tightly with small, round, black seeds. They are not scattered by the explosion as are the discus-shaped seeds of the wisteria. They remain cradled within the holders. Gradually, the contraction of the sides of the pod segments will force them out and distribute them over the ground.

On these November days, seeds are everywhere. I see the watery seed capsules of the green arrow arum, shaped like gourds and almost as long as my hand, breaking open along the shallow edge of the swamp stream. Tan islands of cattail fluff drift on the sluggish current, thousands of tiny seeds riding together. In this all-important matter of seeds, nature never trusts to luck. She banks on the average, the long run, not on the few or the individual. She covers the swamp with floating, parachuted, windborne seeds. A thousand seeds to attain one seedling—that is her method. The chances may be against any one individual seed. But the odds are in favor of the average and the species survives.

NOVEMBER · 9

CATKINS OF RAIN To the garden about 8 A.M. after a night of rain. The trees are drenched and the air is raw. But all down the slope there is rare and shining beauty. The black twigs of the wild cherries are adorned with glowing drops of water. Each drop has collected where a small bud juts from the stem. Thus the spaced drops have the appearance of pussywillows—pussywillows of rain.

NOVEMBER · 10

NESTS IN NOVEMBER Falling leaves have revealed secrets hidden behind the summer foliage. Now I know where the crows nested in the tupelos across the swamp. Now I know from which upper branch the oriole's nest was slung in one of the garden maples. These nests, appearing one by one, have set my mind running back over some of the unusual places I recall where birds have nested and raised their broods.

There was an Eastern phoebe that brought grass and hair and moss and formed her nest on the top of an unused lantern. A ruby-throated hummingbird once attached the delicate little thimble of its nest to the top of an electric light socket. And some time ago, when a bell at a Catholic church, in Troy, New York, began tolling steadily, it was discovered that a bird's nest had been built across the relay points in the control box, closing the circuit and starting the bell ringing.

I remember a killdeer that made its nest beside the tracks of a busy eastern switchyard where freight cars and engines rolled over the rails almost within inches of the eggs. Even more hazardous was the site chosen by a pair of nesting house wrens in a freight yard at Minneapolis, Minnesota. They began carrying sticks into a hole in a hollow wheel on a disabled car that had been shunted on a side track. About the time the nest was done, another derelict was rolled onto the siding. The jolt as it coupled with the first car was sufficient to roll the wheel and turn the nest upside down. In a few days, the wrens were back at work again but, once more, when the nest was completed, a third car was shunted onto the siding and the second nest was inverted and ruined. This time the birds chose another location. A different pair of house wrens lived most dangerously of all. They constructed their nest in a basket, but it was a basket that was filled with sticks of dynamite!

The Rocky Mountain naturalist, Enos A. Mills, once fell from a tree and in so doing his hat caught upsidedown in the crotch of a limb. Mills was injured temporarily and his hat remained in the tree. The next spring a pair of robins built their nest in it. On other occasions, robins have constructed their mud nests in the bulging pocket of a discarded coat, on a tenement fire escape and at the top of a high swinging boom crane.

Swallows, and the curious places they nested, were of special interest to Gilbert White. In his *The Natural History of Selborne*, he recalls a pair that first attached their nest to pruning shears stuck in the boards of an outhouse wall on a Hampshire farm. Then they nested in the mouth of an old conch shell and, finally, in the most curious place of all, on the body of a dried owl that had been shot and nailed up to the wall of a barn.

Eighteen years ago, I was driving toward Washington, D.C., through Franklin County, in southern Pennsylvania, when I pulled up beside a farmyard. A straw stack near the barn was riddled with holes. It looked like a clay bank filled with nesting swallows. The holes were made by English sparrows and their nests were inside the stack. That was in 1935. Never since, anywhere in the United States, have I seen these familiar birds nesting in this manner. Apparently, it is a local habit. But it is a habit that has persisted. For I was interested to note, in opening the June–July 1953 *Nature Magazine*, a photograph of a riddled straw stack inhabited by English sparrows. It had been taken, nearly twenty years later, in this same Franklin County. Here the sparrows have reversed the usual procedure. Instead of carrying straw for their nests, they have brought their nests to the straw!

NOVEMBER • 11

NATURE'S MACHINE How smoothly nature's vast machine whirs on with all the big and little cogs revolving in their places! Each seed and bird and flower and fly, in its apparently haphazard existence, plays its part in the output of the seasons.

NOVEMBER • 12

LEAF FALL On this still, chill November day, I see one of the last of the apple leaves let go its hold and drift downward through the motionless air. No slightest breeze has stirred it. No sudden strain accounts for its fall. Through the action of a special mechanism at the base of its stem, it has joined the windrows of countless leaves that have been shed before it.

Once the work of a leaf is ended, it becomes a liability to the tree. If all deciduous leaves had to be torn away by winter gales, many twigs and branches would go with them. So most trees, with the exception of oaks and beeches and a few others, shed their leaves early. A special separating layer is contained in the cushion, or enlargement, at the base of the stem. This swelling, where leaf stem and twig meet, is, in itself, a marvelous mechanism. It is an organ of movement. It tilts the leaf, lets it droop, adjusts its surface to catch the light. All this is accomplished by the increase or decrease of water in the cells of the cushion. These cells have the power of absorbing water from neighboring cells or of giving it out. When the cells on one side of the cushion are swollen with water, the leaf stem is turned toward the other side. Thus, all summer long, the swelling and contraction of cells within the cushion keep adjusting the position of the leaf.

When fall comes, the separating layer forms. Gradually three tiers of very thin cells extend across the stem of the leaf in the cushion. They develop from the outside toward the center. This separating layer not only shears off the stem, as cleanly as though it were severed with a knife, but it contains a mechanism for developing a positive thrust that pushes the leaf free from the stem. This is accomplished through a breakdown of the middle tier of cells. Their walls seem to be converted into a kind of watery mucilage. At the same time, the cells of the two neighboring tiers begin to bulge. Pressing against each other, they push apart and thus force the leaf from the twig at the separating layer. This not infrequently happens on a quiet day and a leaf, like the apple leaf I have just seen drifting downward, falls to the ground in perfectly still air.

Sometimes in autumn, a sudden drop in temperature will produce a thin sheet of ice extending through the stem in the separating layer. The expansion of the ice forces the separating layer apart but the leaf does not fall because the ice layer holds the stem together. After sunrise, the ice melts and, all at once, a shower of falling leaves drops to the ground below.

NOVEMBER • 13

THE AUTUMN RAINS All day long, cold rain pours down on the streaming earth from a soggy sky. These are the autumn rains. In the afternoon, I go to the dentist. It seems to me that I squirm about more than usual during the drilling. I remark on this and the dentist tells me that he finds people are always on edge on rainy days. Our nervous systems are like barometers that are affected by unsettled weather. Once, in New York, a taxicab driver volunteered the information that he was always tense and keyed up on days when it rained. On a farm, cows and horses are noticeably restless before a storm. And those who have studied shorebirds have observed that the flocks are wilder and more difficult to approach on dark and disagreeable days than when the weather is sunny and settled.

NOVEMBER • 14

DEATH OF A BEETLE I watch a praying mantis moving slowly among the rose bushes in the chill of midmorning today. Its feet advance as though they were clad in lead boots. Cold has slowed it down. In these days of lowered mercury, all the insects still alive move in slow motion.

Thinking of this, I begin to recall dramatic instances of another kind of slowing down in the insect world. This is produced, not by cold, but by a scarcity of nourishment. Before the days of kiln-dried lumber, when timbers were seasoned in the open air, a beetle grub sometimes lived for decades in furniture or porch posts. It had been living a normal existence within the living tree when it was felled. Missed by the cutting operations in the sawmill, it remained in the bit of timber, feeding on a diet so reduced in nourishment that all its life processes were slowed down. The successive stages of its development were immensely prolonged. Thus, from wood that had been in use for twenty, thirty, forty years, suddenly, to the amazement of the householder, there would appear an adult beetle. Henry Thoreau refers to such an instance in *Walden*. There are authenticated cases in which adult wood beetles have reached the end of their slow development and have appeared from wood that has been in use for nearly half a century.

The oddest instance of this kind of which I know was related to me, some years ago, by a man who worked in a chemical factory. The acid vats there had sides of heavy planking. One day, a stream of deadly acid began pouring from a small hole in one of the planks. Investigation revealed that a beetle grub, during its long, slow development, had tunneled through the planking into the acid. It had died instantly. But its tunnel had become the conduit through which the liquid was making its escape.

NOVEMBER • 15

DRY LEAVES Every scampering squirrel, these days, creates a tumult of small dry sounds as it rushes over the carpet of fallen leaves. Now, in the woods, the stalker and the stalked mark their movements with sound. They leave auditory trails under the trees.

NOVEMBER • 16

CEDAR BERRIES On this dull November day, we drive to the cedar woods where, only six months ago, we heard the woodcock sing. The dark shafts of all the cedars are enveloped in a powder blue haze of berries. Below the trees, the mossy earth is sprinkled with the small, round, fallen fruit. We see three winter robins flutter from tree to tree. They are birds that, in all likelihood, will remain in the cedar tract till spring. Along the coast of New England and on Long Island, overwintering robins most often stay among cedar trees. The light blue berries form their principal source of food, supplemented by the gray, waxy fruit of the bayberry bushes and the red seed clumps of the sumac.

NOVEMBER • 17

THE RIM OF EVOLUTION Now that the leaves are off the wild cherries, I can see little varnished, chocolate-colored bands encircling some of the twigs. These are the eggs of the tent caterpillars, massed around their supports and covered with a waterproof, weatherproof coating that will keep them intact until spring. Each of these bracelets contains about 300 eggs.

Some years ago, at the Boyce Thompson Institute for Plant Research, two American scientists, Florence Flemion and Albert Hartzell conducted a series of interesting experiments with these eggs of *Malacosoma americana*. They broke off more than 200 wild cherry twigs holding the egg masses, early in October, divided them into lots of four and placed each quartet of twigs in a different glass jar. They were then subjected to various constant temperatures for periods of from two to forty-five weeks. The results showed the part that winter cold plays in the life of the tent caterpillar. No eggs at all hatched from the twigs kept throughout the winter at room temperatures. And those subjected to freezing temperature the longest, hatched most rapidly when brought into room temperature. Not only do the eggs have to have cold; the more they get in winter at our latitude the quicker they hatch in spring.

One spring, as I previously noted, a sudden drop in temperature after the caterpillars had hatched nearly wiped out the *Malacosoma americana* population in our region. It was years before they built up to normal strength again. I have also noticed that the tent caterpillars among my wild cherries tend to decrease when a series of abnormally mild winters occur. Then they begin to build up again, even if the warm winters continue. Evolution utilizes the fringes. It runs on its rim, so to speak. The great bulk of a species requires certain conditions to exist. But there are always some with a greater tolerance for other conditions—say, some of the tent caterpillars that do better with less cold in winter. If the climate changes, those with this tolerance increase and in the end become the bulk of the population. If the opposite occurs, the few that thrive best in greater cold will have the advantage, will multiply and become dominant. Thus, running on its rim as conditions alter this way and that, evolution is able to maintain a species in spite of endless and erratic variations in the conditions under which it lives.

NOVEMBER · 18

THE VOICES OF THE GRASS I walk about amid the desolation of my garden at 2 P.M. Green is gone from the swampside grass clumps. Yellow stems and lifeless leaves rub together in the changing wind. No longer is there the soft, sibilant, swishing sound of wind in the green fountains of the summer clumps. The voice of the grass has changed. It is thin and dry and harsh and filled with the murmur of complaint. At times, when the leaves rub together in little eddies of air, the sudden sound resembles the dry flutter of locust wings.

Yellows and browns and russets and golds stretch across the valley of the swamp. As I stand beside the dead sunflowers, I hear goldfinches. A band of thirty or more comes trooping from the east, over the stretch of lowland, to alight in the trees of the hillside. For a time, they are all around me in the upper branches, bright little birds with bright little cries. They seem doubly alive in the midst of ebbing life on this November day.

NOVEMBER · 19

PETS Watch the pets and you know the people. A pet that is unafraid is a high compliment to a household.

NOVEMBER · 20

THE SCATTERED BIRDS A few years ago, one of the last bits of wild woodland in our locality was felled for the construction of a subdivision. Would the birds that nested there leave the region? We have watched to see; the results have been interesting. Apparently, a good share of the birds have remained, scattering out to isolated trees and small backyard tangles. Since the felling of the woods, we have had nesting birds near by that were unknown before. These, no doubt, are the old birds, the ones that originally nested in the woodland. They return to their former nesting ground—to use Ernst Mayr's apt description—as though they were attached to it by a rubber band. Once a bird begins nesting in an area, it tends to continue to nest there as long as possible, no matter how many changes occur. It is the old birds that come back, the young birds that scatter. So it seems likely that, because of the reduced facilities for nesting, when these established birds are gone, fewer other birds will come to take their places.

NOVEMBER · 21

FIRST FREEZE When I went to bed last night, a glittering, brittle full moon shone in the cold sky. The thermometer stood at 33 degrees F. and the mercury was still descending. It was the night of the first freeze.

This morning, the hillside is bleak and still. An edging of ice, like frail lacework, runs around the quiet bays of the swamp stream. In this night of cold, only by a miracle could any insect escape freezing. The hush along the winding garden paths is complete. The show is over. The stage is bare. And this passing of the insects—those varied, interesting, colorful creatures of summer days—who is there to regret it? So very few! Like Lucy in Wordsworth's poem, they dwell among "the untrodden ways" and their very existence is not suspected by the vast majority of men. I remember once being introduced to the famous Broadway actor Walter Hampden. A mutual friend explained that I was interested in birds. "They are most interesting," the actor agreed. "He is also interested in insects," my friend volunteered. "I'm afraid," Hampden said with a smile, "I couldn't be quite as enthusiastic about *them*!" So it almost invariably is: Birds, yes. Wild flowers, yes. Trees, yes. Insects, no. Yet nothing in all the infinitely varied enchantments of nature exceeds the fascination of the insects, once their forms and colors and habits and abilities are properly seen and understood. And now they are gone, irretrievably gone, and the garden is stripped of all their life. How sad would be November if we had no knowledge of the spring!

W. H. Hudson, in *Nature in Downland*, speaks of the shadow that fell across his mind when he contemplated this sudden autumn end of the insects' "feast and fairy dance of life." "But," he adds, "it is like a shadow on the earth on a day of flying cloud and broken sunshine that is quickly gone. That teeming life of yesterday has indeed vanished from our sight forever; it is nothing now, and its place will know it no more; but extinction came not on it before the seeds of life that is to be were sown—small and abundant as the rust-colored seed of the mullein, that looked like inorganic dust, and was shaken out of its dead cups by the blast and scattered upon the ground. The still earth is full of it. Out of the matted roots of the turf and from the gray soil beneath, innumerable forms of life resembling those that have vanished will spring to light—creatures of a thousand beautiful shapes, lit by brilliant color, intense in their little lives, forever moving in a passionate, swift, fantastic dance. And we shall see it all again, and in seeing renew the old familiar pleasure."

NOVEMBER · 22

DISASTER INVENTIONS The three great disaster inventions, so far as bringing death to wildlife is concerned, have been gunpowder, the modern match and the gasoline engine.

NOVEMBER · 23

PHEASANTS ON AN ANTHILL A correspondent in New Jersey has just written me of his observation of an interesting habit of ring-necked pheasants. On several occasions, he says, he has seen one of these birds come upon an anthill and begin scratching in it. Sometimes it has kept on until it was level with the ground. Hundreds of ants swarmed over its body and, in twisting and sitting down, the bird crushed many of the insects. After it had gone, examination of the spot revealed that the ground was littered with crushed ants. "In my opinion," my informant adds, "the birds crush the ants to get the formic acid on their bodies." This is a new mode of anting. It is one of which I had never heard before.

NOVEMBER • 24

STRENGTH OF A WASP NEST Among the bare branches of one of the wild cherry trees, today, I find a deserted nest of the paper-making wasp, *Polistes pallipes*. It is flat and about four inches across. The hard paper neck that attaches it like a stem to the branch is so vital to the safety of the nest that I begin to wonder how strong it is. It must not only support the weight of the loaded nest but also must meet the side strains of gusts and gales. What is its factor of safety, its excess strength beyond that needed merely to support the weight of the nest? I come back in the afternoon with delicate spring scales and make tests. I find that a sidewise pull of one pound, nearly a hundred times the weight of the paper shell, is needed to tear the nest loose from its anchorage on the limb. And, to detach the neck from the nest proper, I discover it is necessary to exert a straight pull of six pounds—half a thousand times the weight of the empty nest.

NOVEMBER • 25

FROST MAN A swift drop in temperature came early last night and this morning we find the kitchen windows bordered with the beauty of the first frost paintings of the year. Fern forests, palm oases, submarine scenes filled with floating seaweed—these and a thousand other forms of crystalline art are represented, each year, in the frost patterns on winter windowpanes. Creations of congealing moisture, they represent one of the most ephemeral forms of art. They reach a swift climax of beauty, remain but a few hours, then are gone forever.

Whenever I see frost patterns on glass, I recall a memorable winter day spent with Dr. Lester W. Sharp, of the Cornell University botany department. For more than a decade, Dr. Sharp has been preserving on photographic film the fleeting beauty that cold and moisture have combined to produce on his windowpanes. These negatives form one of the world's outstanding collections of the kind.

As we looked over picture after picture, he told me he has never encountered two formations exactly alike. They vary like the patterns of snowflakes. But certain general types of frost designs can be traced to the conditions under which they are formed. Palmlike patterns, for instance, are most likely to occur when a uniform film of water coats the windowpane. A more gradual deposit of moisture from the air, on the other hand, creates delicate, feathery frost designs.

Once Dr. Sharp timed the formation of one of these crystalline ferns. He found it was growing at the rate of about half an inch a minute. When such a fern comes in contact with another fern that has crossed its path, it stops growing instantly. All of the photographs in Dr. Sharp's collection have been made with the same simple equipment, a forty-year-old view camera that stands on its tripod, loaded and ready, during the winter months. It is used most often at daybreak. The crystalline patterns are at their best during the early morning hours. By nine o'clock, their ephemeral beauty is already fading.

NOVEMBER · 26

NOVEMBER GALE Last night, scudding clouds raced across the face of the brilliant half-moon. Its light went on and off. It seemed successively to pop from hiding and to duck back in again. And, seen in silhouette against the sky, all the bare branches swayed and rubbed together in the wind.

This morning, we woke to greater winds. A November gale, hurling itself on the land from the sea, was increasing in violence. All day, it has battered the coast. We hear the storm roaring through the trees and smashing against every obstruction that rises above the ground. Two limbs from the backyard maple have crashed into the driveway and one of our fences is listing far to one side. A whole section of shingles is gone from a neighbor's roof. With the high tide, the wind has sent salt water flooding all across the sea meadows. The roaring of the great combers on the outer beaches, miles away, is loud on the gale winds. Battened down, we are sitting out the storm. Electricity is off. After dark, as I write this by the light of a candle, the slashing of the rain, the mauling of the wind, the rattling of the windows, the shrieking of the gusts, are part of the night. What is happening, I wonder, beside the swamp, at Milburn Pond, among the old trees of my Insect Garden?

NOVEMBER · 27

IN THE WAKE OF THE WIND Looking from my bedroom window, in the watered-milk light of early dawn, I see the trees are still. A kind of exhausted quiet has fallen over the out-of-doors. The great November gale has blown itself out in the night. Before breakfast, I walk to the garden hillside. Along the way, wires and trees are down and television aerials are tilted at precarious angles. I find the whole slope, in the wake of the wind, strewn with wreckage.

One of the red cedar trees beside the swamp lies prostrate. Twigs and branches are scattered everywhere. In three of the old apple trees, the whole tops have split away. The upper stubs of the trunks rise starkly with shining, yellow-tinted sapwood enclosing the dark-chocolate, ant-riddled interior. Here, again, the carpenter ants have destroyed themselves. Every broken tree has been weakened by the mining of the insects. All across the lowland, the cattails are flattened and broken, trampled by the gale. And, at Milburn Pond, sassafras and tupelo, dogwood and sweet pepper bush have been raked and ripped by the wind. Even the tough and tenacious cat-briars have fallen victim to the gusts.

Everywhere I walk, there is evidence of catastrophe and destruction. But there is something exciting and stimulating in this wreckage of elemental battle. Some trees have weathered the storm; some have fallen in the gusts. But all have been subjected to an emergency that trees were designed to meet. Their disaster, in the midst of the superlative fury of the elements, was no maiming by vandals, no death by fire through some carelessly tossed cigarette stub, no wanton hacking down. Theirs was a spectacular but a natural end. It is with such thoughts running through my mind that I finish my census of destruction and turn toward home.

NOVEMBER · 28

SCRATCH WITH THE HENS Of all the varied attitudes of mind, it seems to me the one that brings the most enjoyment in the out-of-doors is that quirk of imagination that lets you participate in fancy in the lives of the creatures you meet. John Burroughs expressed this outlook when he wrote to a friend: "My thoughts go and scratch with the hens, they nip the new grass with the geese, they follow the wild ducks northward." And so did John Keats when he said: "If a sparrow comes before my window, I take part in its existence and pick about in the gravel."

NOVEMBER · 29

MANTIS EGGS On this day, I range far over weed lots and bushy tracts in search of the egg cases of the praying mantis. Each tan-colored ball of hardened froth is about the size and shape of an English walnut. It holds from 125 to 350 eggs. Attached to bare twigs and the stems of weeds, the egg cases are difficult to see. Their tan coloring merges with that of the dry weeds. When first I began hunting these oöthecas, I found I could train my eyes to see them if I scattered several around in weeds and bushes and then pretended to find them. Each time I pounced upon one of these planted egg cases, I used to chuckle at the remembrance of a somewhat similar ruse employed by Australian aborigines. James G. Frazer, in *The Golden Bough*, relates how one of these superstitious tribesmen would set a snare, then walk casually around, let himself get caught and cry: "Hola! I believe I'm caught!" This was supposed to make animals come and do likewise!

In my wanderings today, I am interested to observe the supports to which the insects have attached their oöthecas. One winter, some years ago, I examined 258 *Tenodera sinensis* egg cases, noting the supports to which they were attached. The results, which were published in *The Bulletin of the Brooklyn Entomological Society*, showed that bayberry twigs are preferred in this locality. Seventy-six of the oöthecas were on such twigs. Other supports, in the order of their preference, were: goldenrod, fifty; privet, twenty-nine; sumac, twenty-five; wild cherry, twenty-two; blackberry, twenty; honeysuckle, eighteen; grass stems, three; cat briars, two; miscellaneous, thirteen.

Another interesting thing about these egg cases I discovered almost by accident. In the winter of 1943, I collected a large number of which all but twelve hatched in the spring. This dozen oöthecas, after being placed in a Mason jar, were put aside and forgotten. The second spring, in 1944, three of them hatched. This is the only record, so far as I know, of delayed hatching in the eggs of the mantis. The hatchings of the second spring, it might be added, were comparatively small. The total number for the three egg cases was 112 young mantes.

NOVEMBER · 30

WHAT DO ANIMALS KNOW? On this last afternoon in November, I see Silver sitting in the backyard gazing up at the white clouds drifting by. There is nothing there that I can see of special interest to him. I remember once, years ago, seeing a horse standing in a pasture staring intently, minute after minute, at the sunset clouds. Do animals ever appreciate natural beauty? We see little evidence of it. But who can say? Thomas Hardy told a friend that he often looked at a kitten that followed him about and wondered what it knew that he did not know. Who can tell by looking at a man, standing silent under the night stars, what thoughts are running through his mind? It may well be that animals, too, take a silent delight in their surroundings. We do not know. But we do know that we human beings, to a greater or lesser degree, possess the priceless gift of such appreciation. Each of us can look back to some early hour when the sunset suddenly became a glory in our mind. It was like the first day of *our* creation.

CHAPTER TWELVE

December

DECEMBER • 1

DECEMBER DAWN Gray light from a gray sky spreads over the earth in the slow dawn of this first day of the last month of the year. The air is raw; the thermometer stands at 31. Bundled in a sweater and a leather jacket, I walk about the garden hillside amid the wreckage of the storm. Across the swamp, the overwintering blackbirds, nearly one hundred of them, swirl upward from the wind-slanted phragmites. I watch them sweep and circle. Then they wing their way toward the east to feed in the open fields. A gray squirrel, with a mouthful of dry leaves, sprints up the trunk of a maple tree. He is hurrying to add another blanket to his bed. As I stand there, a loud and sudden rattle comes from above the swamp and I am surprised to see a kingfisher so late in the season bounding along above the winding stream. I work my way back along the edge of the swamp. A song sparrow, silent now except for an occasional slight lisping call, flits from bush to bush ahead of me.

DECEMBER · 2

ROSES BY ANOTHER NAME "Books of natural history," Thoreau wrote, "make the most cheerful winter reading." On a dark December day like this, there is a special pleasure in thumbing through books of botany. They bring back the perfumes and colors of the flowers of spring. There is fun, too, in noticing relationships in the plant families. There are, for example, many roses by another name. Strawberries are roses. Blackberries are roses. Raspberries are roses. At least, they all are members of the great rose family. So are spirea and hardhack bushes.

Wandering among the pages of a botany, you come upon striking contrasts in related plants and you occasionally discover black sheep and eminent citizens side by side. Consider the potato. Potatoes are the mainstay of the diet of millions of persons. They have made possible the concentrated populations of a number of European countries. Yet in the same family are the deadly nightshade and the troublesome Jimson weed. The petunias that bloom in dooryards and window boxes are related to the potatoes. So are the horse nettles and tobacco. So are the ground cherries and all the tomatoes that form one of the great garden crops of America.

Even such contrasts afforded by the potato group are less striking than those found in the pea family. Of all the botanical clans I know, it embraces the widest range and the greatest contrasts. It is not surprising to find that peanuts are in the pea family. But who would expect locust trees? Or the furze and gorse of English literature? Or the loco weed of the West that poisons horses and cattle and is familiar to every reader of cowboy fiction?

Long before the coming of the White Man, the American Indians were using the groundnut as a staple food. Growing in strings in the soil, these tubers form an excellent source of starch. The groundnut was the potato of the Indian. Yet, botanically it is not allied to the potato at all. Instead, it is a relative of the furze and the locust tree—a member of the pea tribe.

Ground plums, as well as groundnuts, belong to this family. So do the showy tick-trefoil or beggar's-lice and all the alfalfa and vetch and clover that cover vast stretches of farmland and form an important crop in American agriculture. The Kentucky yellowwood tree belongs to the clan as does the American licorice root and the lupine plants that add their colorful flowers to the beauty of gardens and fields and mountain meadows. Thus, in one strangely assorted botanical family, we have everything from peanuts to beggar's-lice, from loco weed to licorice root, from gorse to lupines. And all are related to the familiar beans and peas of our gardens.

DECEMBER • 3

RUSE OF AN INJURED BIRD About a week ago, one of the bluejays in our backyard flew into something or was injured by a predator. Its right wing is weak. The bird can fly on a long downward slant but it cannot rise into the air. Day after day, we have watched this jay resort to the same intelligent stratagem. It feeds close to a snowball bush at the back of the yard. As soon as it has finished eating, it hops into the bush and works upward among the branches until it reaches the top. Here, it jumps across to the lower limb of a maple tree. Hopping from branch to branch, it ascends to the topmost limb. After resting a while, it launches itself out into the air and flies on a long downward slant to a group of trees in another yard. Later, it climbs to the top of those trees and toboggans back to the maple. Thus it travels between its feeding place and the tangle beneath the trees in the other yard, where it hides from enemies, without running the danger of hopping over the ground. Its wing is strengthening rapidly and in a few days it will be able to fly normally. Its action recalls the similar stratagem of the cicada-killer wasp. This large yellow-and-black insect stocks its tunnels with cicadas. When they are too heavy for it to lift directly into the air, the wasp will drag them up the trunk of a tree and to the tip of an upper branch. Launching out from this height with its burden, it is able to cover considerable distance on a downward slant in the direction of its burrow.

DECEMBER · 4

WHAT WE SEE We see only what we appreciate and we appreciate only what we understand. So it follows that we see hardly a thing we have not already seen in our minds.

DECEMBER · 5

LEOPOLD'S RABBITS Grasp any thread in the web of life and you find it joined with a thousand other threads. Leopold's rabbits are a case in point.

Today, while the cold wind has snuffed and pawed along my study windowpanes, I have been reading under my desk lamp, as snug as any squirrel in its leaves or any white-footed mouse in its made-over bird's nest. And one of the things I have been reading is an account by the late Aldo Leopold of the relationship between rabbits, ladyslippers, deer and grouse in the Wisconsin woods.

Between the years 1932 and 1935, he found the rabbits especially abundant. They nibbled down an unobtrusive little bush, the bog-birch. Starving deer live on the tips of these bushes and, in times of prolonged blizzards, the sharptailed grouse eat the buds and stay plump. So the fate of grouse and deer was linked to the fate of the bog-birch. So, in a different way, was the increase or decrease of ladyslippers. When the rabbits ate back the bushes and more sunlight got to the floor of the woods, the ladyslippers increased. Then, during the years 1936 and 1937, a crash in population decimated the rabbits. The bog-birch regained its former density, deer and grouse were provided with ample emergency food, but the ladyslippers, cut off from the sun, decreased in consequence.

I remember William Vogt telling me of an even more unexpected relationship in the web of life: how the hoof-and-mouth disease benefits songbirds in Argentina. This disease restricts the exportation of beef from the country. As a result, there is a plentiful supply of meat in Argentina. Should the disease be conquered and large quantities of beef suddenly be exported from the nation, the meat-hungry citizens of Argentina, where bird life is held in such small regard, would soon turn to the slaughter for food of many species. Thus, by a tenuous thread, the welfare of New England bobolinks that overwinter in Argentina is dependant upon the persistence of a livestock disease!

DECEMBER · 6

MERRY-GO-ROUND Hardly had Nellie thrown some crusts of bread into the backyard this morning, before the starlings and English sparrows came flying. One sparrow grasped a part of a crust and tried to take wing. The bread was too heavy. As it clung to the fragment, a starling walked up and grabbed the other end. The sparrow clung to its prize. The starling jerked. The sparrow hung on. In the end, the starling swung the bread, with the sparrow hanging to it, around almost in a half circle. It suggested a hammer thrower at a track meet. This was too much for the smaller bird. It flew away and the starling ate the bread.

DECEMBER · 7

PLANTS THAT DISSOLVE STONE The cedar woods today are so still I can hear the cawing of a faraway crow and the calling of a bluejay beyond the fields where the woodcock rose on its song flights of spring. Here, among the bare bushes and leafless trees, the green lichens come into their own. The summer foliage engulfs and hides them. Now the foliage is gone and only the green of the mosses and the catbriars is left to complete.

I see one lichen spreading across the surface of a small stone and it reminds me of the amazing relationship between rocks and lichens. Some of these plants, so frail they can be crushed between a thumb and forefinger, are able to dissolve granite. They produce rock-etching acids that eat out tiny pits and depressions in the stone, thus enabling the threads of the lichen to strengthen their hold on boulders and cliffs.

In the laboratories of research chemists, I am told, nearly 150 lichen acids have been isolated. They are fluids that are found nowhere else in the world. It is the crystalline form of these acids, incidentally, that is responsible for the red and orange and silver and yellow hues that make brilliant the most colorful species of the lichens.

DECEMBER · 8

SPECIALISTS Today I had lunch in the city with two scientists, a botanist and an ichthyologist. The botanist said he never kept a garden and the ichthyologist said he never went fishing.

DECEMBER · 9

THE HIDDEN SEED The smell of coming snow is in the air. All the birds are feeding ravenously—a sign of bad weather. As I walk along the hillside under the heavy sky, a little band of myrtle warblers flies before me. The birds are feeding on goldenrod seeds. They dart ahead and I see them land amid the dry, broad-leaved grass of the Katydid Jungles, the corn panic grass, *Panicum clandestinum*. The "clandestinum" of this scientific name has special significance. Some of the seeds of the grass are hidden. One bird seems to have discovered the secret of the panic grass—the myrtle warbler. It is the only one I know that tears open the plant to get at this cache of hidden food.

DECEMBER · 10

KEEN EARS Snow in the night, the first snow of the winter. The flakes are granular and fine and only a powdering lies on the swamp trail when I walk along it after breakfast. I stop beside a little overarching bower formed by a grass clump. The grass is yellow and dry but it still shelters the spot below, keeping it clear of snow. There, a cottontail had its form when the clump was green. I remember coming upon it in the heat of an August afternoon when the animal was sound asleep. I could see only its nose and part of an ear and a tightly shut eye through the thin slits in the curtain of grass. I watched from a distance through my field glasses. Although the cottontail was asleep, its nose kept twitching, wrinkling and sniffing. A rabbit sleeps with its nose awake.

Similarly a cat sleeps with its ears awake. At the first sound of a can opener in action, Silver is on his feet. We once had a cat, Frosty, that slept curled up under a bush in a far corner of the yard. He would come racing at the first stroke of a knife being whetted on a stone. William T. Davis, the authority on American cicadas, once told me of an amusing experience with a keen-eared cat in Virginia. He was staying on a farm making a collection of cicada nymphs as they emerged from the ground. The cat had developed an appetite for cicadas. Even at a distance, it would catch the slight sounds made by the legs of a nymph climbing through the grass and would run in that direction. The collector soon depended on the cat to discover the insects and the two had many a race to see which would get to the cicada first.

Cats are especially keen-eared in the matter of high-pitched sounds. But the mammal that undoubtedly exceeds all others in this respect is the bat. Its life depends upon catching echoes of the shrill sounds it emits as it flies through the dark. When the Swedish scientist, Dr. Olaf Ryberg, visited the United States a few years ago, he told me of the remarkable feats of hearing performed by bats he kept in captivity. They apparently could hear a fly cleaning its wings or rubbing its legs together. Immediately the ears of the bats would be erected and pointed in the direction of the insect. Then the bat would dart in that direction and snap up the fly its ears had first discovered.

DECEMBER · 11

COWPATHS AND FAIRYLANDS Mentioning William T. Davis has brought another memory of that noble and kindly man who lived his whole life close to the spot where he was born and who found his pleasures in simple things near by. He once showed me some of the unpublished things he had written. I remember two eloquent sentences that express the whole outlook of his life. "There is no need of a faraway fairyland," he wrote, "for the earth is a mystery before us. The cowpaths lead to mysterious fields."

DECEMBER · 12

THE GREAT GUPPY MYSTERY The postman, a few days ago, left a first-class natural history mystery in my mailbox. It is the mystery of the missing guppies. It comes from a bewildered reader in Philadelphia.

"I used to have fifteen guppies," he writes. "I kept them in a regular glass fish bowl. The bowl got broken accidentally and I put the fish in a basin. About three days later, one of the little fish disappeared. I could not find any remains of it anywhere. I thought it had died and the other guppies had eaten it up. Then another disappeared and another. This kept on until there were only three guppies left. Then there were only two; then only one."

Well, up to this point, the answer seems simple. Guppies are known to eat each other up. This might go on until there was only one left. But . . .

"Two days later, the last guppy disappeared without leaving a trace behind!"

Maybe there was a cat in the house. But . . .

"I don't keep a cat. I don't have any other pets in the house. What I want to know is: What became of the guppies? I wonder if mice could have carried them away. I do not take much stock in this idea because I do not think a mouse could swim around and catch a guppy. This has me baffled. Can you suggest any solution to this mystery of the missing fish?"

That is a question I have been passing on to my friends since I have received this letter. Some think it must be rats. But Dr. Myron Gordon, of the American Museum of Natural History, who has been carrying on researches with these midget fish and has watched thousands of them, has a different explanation. At the temperature at which the water is kept in guppy tanks, he says, bacteria will completely destroy the body of a fish in a surprisingly short time. Within the space of twenty-four hours, everything will have disappeared except the tiny skeleton which is likely to escape notice. Perhaps the guppies died one by one and the bacteria swiftly consumed their bodies.

Another ingenious solution proposed relates to the fact the fish were in a basin. Its sloping sides, it is pointed out, might have acted in the manner of the slanting boards Oriental fishermen put over the side of their boats on moonlit nights. The boards are highly polished and reflect the moonlight. The fish dash for them and slide up the smooth boards and into the boat. This, it is suggested, is what happened to the guppies. They dashed up the side of the basin and onto the floor. By flopping around and gathering lint on their sticky sides they become camouflaged and invisible.

And that is where the mystery of the missing guppies stands at this moment.

DECEMBER · 13

COLLISION This morning something happened in the backyard that I have never seen occur there before. It was a collision. I had tossed some peanuts from the back door onto the frozen ground. A gray squirrel ran down the trunk of a maple tree and headed for one of the nuts. At the same time, a bluejay, that had been hopping about among the limbs of a cedar tree, flew toward the same nut. Neither seemed to see the other until it was too late. They arrived at the nut at the same time and collided. The jay flew up with a shriek. The squirrel rushed away in a panic. It was minutes before they came down from their respective trees. This time they went to separate peanuts.

DECEMBER · 14

SCENT The scent a bloodhound catches in your trail is more precious than any perfume. It is unique. It is individual. It is you.

DECEMBER · 15

THE SLEEPING ANTS Not far from here, a friend of ours moved into a new house. When he unpacked a box of odds and ends, he discovered that a carpenter-ant colony had moved with him. The box held such things as photographs and teacups wrapped in newspapers. It had been packed in the fall and stored in the garage. There, the black ants had picked it as a snug place for hibernation. Some of the stacked teacups were half full of the massed insects. One framed photograph was almost entirely obscured by a layer of ants that had crept into the narrow space between the glass and the picture. All of the insects were motionless, torpid, lost in the long dormancy of their winter sleep.

DECEMBER · 16

BROWN CREEPER Hitching upward along the bark of an apple tree this morning, a brown creeper hunted for hibernating insects and insect eggs. As I watched it, I remembered an old description of the bird: a mouse in a salt-and-pepper suit. Its coloring so perfectly fits its surroundings that when it pauses, its stiff tail feathers braced against the trunk, it seems to disappear. However, this welcome little bird of winter is one of the easiest to recognize. Its methodical habits reveal its identity. It always works from the base of a tree upward, descending from high on one tree trunk to the bottom of the next tree trunk. This sets it apart from other birds of the winter woods.

Several times this morning, I watched the brown creeper change from one tree to another. Each time, it alighted about two feet from the ground and began ascending. It moved upward around and around the trunk as though on spiral stairs. Twice it hitched itself woodpecker-wise straight up the trunk for a foot or so. Then it resumed its ascending spiral again. Through my glasses, I could see its bright little eyes peering closely at the bark crevices before it and from time to time it would pause and its downcurving needle of a bill would probe for hidden food.

At rather rare intervals, I heard the few squeaky, high-pitched notes that make up its call. Only in the Canadian and Transition Zones, where the bird nests, is the real voice of the brown creeper known. There, during the spring breeding season, the male has a warbling, beautiful song. I have never heard it. But William Brewster, the famous Cambridge ornithologist, describes this performance amid the northern firs and spruces as an exquisitely pure and tender song of four notes, the last abruptly falling, "but dying away in an indescribably plaintive cadence, like the soft sigh of the wind among the pine boughs."

Nobody knows how many insects are destroyed by the winter feeding of the creepers. Pupae, eggs, adults in hibernation, as well as the eggs and adults of spiders, are consumed in quantities. Over and over again, the little birds comb the bark of the trees. Their long claws give them a secure hold on their support. And the stiff, pointed feathers of their tails brace them in position. By form and inclination, they are fitted to find their food only when climbing.

In fact, when there are no trees at hand, the brown creeper will sometimes hitch itself up other and strange supports. These birds have been observed climbing up rocks, fence posts, a sand bank and the brick wall of a city house. Witmer Stone tells of seeing a brown creeper laboriously ascending an iron standpipe near the lighthouse at Cape May, New Jersey. It was following the rivet heads that rose in a straight line from bottom to top, feeding on the insect pupae that were secreted in the cracks between the rivets and the sheet iron of the structure. But strangest of all these odd supports was one noted on Block Island, Rhode Island. There a brown creeper was seen climbing, in a characteristic series of hitches, up the tail of a cow!

DECEMBER · 17

THE GREASED POLE Somewhere in *A Traveller in Little Things*, W. H. Hudson remarks that "if there is one sweetest thought, one most cherished memory in a man's mind, especially if he be a person of gentle pacific disposition, whose chief desire is to live in peace and amity with all men, it is the thought and recollection of a good fight in which he succeeded in demolishing his adversary. If his fights have been rare adventures and in most cases have gone against him, so much the more will he rejoice in that one victory." Today, we understand his reflection. For, today, we have outwitted a squirrel. The redoubtable Chippy has been humbled and brought low.

A week ago, we installed a bird-feeder at the top of an iron pipe set in the ground outside the living room window. Chippy watched the proceedings with an expectant eye. But the pipe was metal; her claws could not penetrate it; she would not be able to climb it. That was our reasoning. We were sure. We were also wrong. We loaded the feeder with such choice fare as sunflower seeds and retired to let nature take its course. In three minutes it did. We looked out. Chippy sat in the feeder stuffing herself with sunflower seeds. I chased her away and watched. She came back and scrambled up the pipe as though it were the limb of a maple tree. Apparently, little pits and roughened places on the metal were all she needed for a foothold.

The next day, I hammered out a wide cone of tin to form a squirrel-guard and placed it well up on the pipe. Chippy reconnoitered the new impediment from several angles, cocking her head first on one side then on the other. Finally, after several tries, she leaped to the edge, found a clawhold where the tin was fastened together, hung by one foreleg for an instant, pulled herself up and scrabbled over the tin and again sat in the Promised Land of the sunflower seeds.

It looked as though we had suffered one more in a long line of defeats when our next-door neighbor, Percy Verity, provided the ammunition that brought glorious victory. He came over from his garage with a pail and brushed onto the pipe a black coating of waterproof, slippery graphite. Then we retired indoors, rubbing our hands in glee that is sometimes described as unholy. Nor were our expectations unjustified. Chippy came confidently hopping over the lawn. She paused, looked expectantly up at the feeder, gave a great and graceful bound to the pipe and shot downward in a whizzing backward descent. If ever a squirrel registered the human emotions of outrage, mystification and disgust it was Chippy. She looked at the pipe from a distance. She sniffed at it close up. She examined her paws. But she made no further effort to climb the bewitched support. Our triumph was complete. We tossed out a handful of peanuts; we could afford to be magnanimous in victory.

DECEMBER · 18

WINTER BATHING The mercury this morning is in the mid-twenties and a cutting wind blows out of the northwest. Yet as soon as we put out a pan of water in the backyard, starlings are splashing and bathing in it, ruffling their feathers and sending the water flying. This is a thing we notice each winter: of all the birds that visit the yard, the starlings bathe the most during the months of cold. Can it be that they are hotter birds? I turn to my library. A starling, I find, has a temperature of 109 degrees F. An English sparrow runs from 107 to 109. The bluejay exceeds the peak of both at its maximum body temperature. It ranges between 106.2 to 110.2. Yet a bluejay bathes much less frequently than a starling. So the explanation lies elsewhere than in the temperature of the bathing birds.

DECEMBER · 19

WARMTH OF THE FIRE On this dark, cold, closed-in day, I have just read a letter written by Henry Thoreau to his Worcester friend, Harrison Blake. It was set down on a day in December almost a hundred years ago.

"I have just put another stick into my stove—a pretty large mass of white oak," Thoreau wrote. "How many men will do enough this cold winter to pay for the fuel that will be required to warm them? I suppose I have burned up a pretty good-sized tree tonight—and for what? I settled with Mr. Tarbell for it the other day; but that wasn't the final settlement. I got off cheaply from him. At last, one will say, 'Let us see, how much wood did you burn, sir?' And I shall shudder to think that the next question will be, 'What did you do while you were warm?' "

Thirty years ago, I first read those words. As I laid down the book in which I had come across them again, a wide, brown river seemed carrying me along, flowing away before me in lazy, sweeping curves beneath a late June sun. The river was the Ohio and I was rowing the 500 miles between Louisville, Kentucky, and the Mississippi. Those drifting river miles are vivid in my mind on this dark day in December. For somewhere along the way, at a small secondhand bookstore in some forgotten river town, I bought, for five cents, a Haldeman-Julius Little Blue Book: *The Epigrams of Thoreau*. For the rest of the trip, I carried it in a shirt pocket, where it became stained and frayed. During those golden, toasted hours, I used to row for ten or fifteen minutes and then drift on the current and read the succinct thoughts of Thoreau. This was my memorable first introduction to his writings. Of all that I read in that tattered little blue-covered book, this portion of Thoreau's letter to Blake has returned oftenest to my mind in the intervening years. Coming upon it today was like meeting a friend—a thoughtful, clear-eyed friend—of old, adventurous times.

DECEMBER • 20

THE WINTER WIND Snow clouds hasten the dusk of this short day. For several hours, the wind from the northwest has been rising. Now, in the twilight, it is sweeping across the island laden with snow. Long scudding clouds of white stream past the lighted windows. Wind-driven flakes scratch and hiss along the panes of glass. I step out into the yard, clad in a leather jacket and with hat pulled low. The sweep of the snow slashes at my bare face and hands. The yard is white and each tree has its northwestern side plastered thick with wind-packed snow. Around the house, drifts are forming. I see through the swirling flakes, where light streams from the kitchen window, that already foundations have been laid for the long, tapering, streamlined V's of snow th morning, will extend downwind behind each small obstruction. The winter wind, on winter's eve, has brought the first heavy snowfall of the year.

DECEMBER • 21

EVENING IS OVER On this, the shortest day of all the 365, I wander over the covered paths of the garden hillside. I wade through the drifts along the swamp edge. I walk over the snow-covered ice among the cattails. The wind is gone. The day is still. The world is decorated with unmarred snow. This is winter with winter beauty everywhere.

Autumn is finally, officially, gone. Like the evening of the day, the fall has been a time of ceaseless alteration. Cold, in the autumn, is overcoming the heat just as darkness, in the evening, is overcoming the light. All around, in recent months, there have been changes in a thousand forms. The days of easy warmth were passing, then past. Birds departed. Threadbare trees lost their final leaves. Nuts fell from the branches. Pumpkins and corn turned yellow in the fields. For animals and men alike, this was the time of harvest. The phantom summer, Indian summer, came and went. The chorus of the insects died away in nightly frosts. Goldenrod tarnished; grass clumps faded from green to yellow. Milkweed pods gaped open and their winged seeds took flight. The windrows of fallen leaves withered, lost their color, merged into one universal brown. Now they are buried beneath the new and seasonal beauty of the snow. Autumn, the evening of the year, is over; winter, the night of the year, has come.

DECEMBER • 22

DARWIN I sometimes wonder what difference it would have made in the reception of Charles Darwin's *The Descent of Man* if it had been called *The Ascent of Man*. Any advertising copy-writer would have told him to change the title. But the rugged honesty, the undeviating devotion to scientific method and thought that were cornerstones of his greatness are reflected in the title he chose.

DECEMBER · 23

THE MOUSE AND THE CHICKADEE Hanging on the suet-feeders, flying away with sunflower seeds, investigating crevices in the bark, a little band of chickadees has filled the old maple with life on this cold winter day. It is chickadee time all across the northern states. The friendliest and most open-mannered of the winter birds, the little black-cap is performing invaluable service through his consumption of insect eggs.

Many a boy has first become interested in birds through some sociable chickadee that came close, hopping from branch to branch, stopping confidingly only a few feet away, cocking its head and peering intently into his face with its bright little eyes. And many a man has remembered such moments in later years and, like Frank M. Chapman, thought of the chickadee as his best friend among the birds. On the morning after the recent night of snow, I came upon half a dozen cheerful, calling chickadees performing gymnastics among the twigs of the apple trees. A troop of these friendly little birds met amid falling snow or among the bare winter trees is a sight that has brought a glow to many a woodland walker's heart.

Among birds, I suppose the black-capped chickadee ranks close to the top in inquisitiveness. It is always investigating, always coming closer to see what is happening. A friend of mine was once standing quietly in the winter woods when a mouse appeared from a hole at the base of a tree, scurried about among the fallen leaves, then darted back into its tunnel again. A chickadee, investigating the bark of a limb overhead, spied the little animal. It followed its every movement. When it disappeared, the chickadee darted down and peered into the entrance of the tunnel. Then it flew away only to return again, hop about the hole, cock its head on one side, peer again and again into the burrow. It even perched at the top of the tunnel entrance and hung upsidedown to get a better look inside.

DECEMBER • 24

MERMAIDS I have just come across the record of an event in northern Scotland, fifty years ago. In 1900, in a remote coastal hamlet in Sutherland, a bedridden old man, more than eighty years of age, declared "in a sworn affidavit in front of his minister" that he had, that day, seen a mermaid in the near-by lochan and that the mermaid had beckoned to him. This account recalls a fact often lost to mind, namely, that the important thing is not who says it, or the conditions under which it is said, but—is it true?

DECEMBER • 25

ROBINS AND HOLLY BERRIES Christmas Day dawns with cold and cloudless skies. But beneath them, countless homes are bright with colored lights, decorated trees and holly wreaths hanging at the doors. Not far from here, such a wreath was the scene of unusual activity during one recent Yuletide season. It hung on a front door under a porch near the cedar woods where robins overwinter. Toward midday, the owner of the house heard a soft, fluttering sound along the glass of his front door. He peered out. Against the snowy landscape beyond, he saw the red breast of a robin. The bird was one of the dwellers in the cedar tract. Flying past, it had caught sight of the holly wreath and had fluttered down to feed on the bright red berries.

DECEMBER • 26

THE AILING GULLS This afternoon, Nellie and I drive to the breakwater at Atlantic Beach to look for eiders. A bitter, salt-laden wind pounds in from the sea. It carries the sand in long, low-blowing lines until the whole beach seems in motion. We struggle against the gritty gusts, taking shelter behind pilings, protecting our glasses inside our jackets, until we reach the great rocks of the breakwater. They are icy with frozen spray. Moving carefully, we work our way out to the shelter of a larger rock. Here we catch our breath. Then we begin sweeping the tumbling, spray-crowned combers with our glasses. We see three loons, some red-breasted mergansers, a few dark, double-crested cormorants, a raft of white-winged scoters—but no eiders. All the birds are riding up and sliding steeply down as the great waves sweep under them.

At the end of twenty minutes, we pick our way inshore along the breakwater. The wind is now at our backs and we notice, for the first time, a herring gull hunched up in the lee of one of the rocks. It apparently is a sick bird finding shelter from the cutting wind. A little farther on, we come upon another, a second immature gull. By the time we have reached the sand, we have discovered four hidden among the rocks. In the shelter of pilings, we see two others. All are immature herring gulls. In addition, we see four or five dead birds lying among the rocks or on the sand. This breakwater apparently forms a sanctuary from the wind sought by all the ailing gulls of the vicinity.

DECEMBER · 27

POSTSCRIPT ON A SQUIRREL At 11:03 this morning, the redoubtable Chippy sat among the sunflower seeds in the feeder at the top of the graphite-covered pipe. I dashed out of the house. Chippy raced down the pipe and bounded away. I examined the graphite. Its magic is gone. It has weathered and hardened. Now it covers the metal like bark on a tree limb, providing an easy ascent to the Elysian fields of the feeder. Paradise Lost was, for Chippy, Paradise Regained. For us, however, there remained a moment of triumph unforgotten, a mental motion picture of the redoubtable Chippy whizzing downward in her one greased-pole descent to earth.

DECEMBER · 28

THE CENTER OF DAYS "Where I am or you are this present day," Walt Whitman wrote, "there is the center of all days." For me, on this dark, twenty-eighth of December, the center of days is the heart of a snowstorm. I walk across the fields in silence, shut in, surrounded by curtains of drifting white. My footfalls are as noiseless as though I walked on cotton batting. The heart of this storm is tranquil. Here there are no wind-hurled flakes, no driving sheets of snow. Here the air is unmoving. The flakes descend slowly, gently, a procession of crystals from the sky.

DECEMBER · 29

BIRD COUNT This is the day of the winter bird count. Each year, the same five square miles are covered intensively by members of the Baldwin Bird Club. Twenty-two of us were out today. The skies were clear and cloudless over the snow-covered earth. It was dawn when I began at Milburn Pond. It was dusk when we all got together to compare notes. During this day, in the five square miles, we had seen more than 4500 individual birds belonging to sixty-seven different species. They ranged from common loons and Canada geese to European widgeon and greater scaup and from marsh hawks and short-eared owls to northern horned larks and slate-colored juncos.

Among them all, the cowbirds presented the most puzzling paradox. For several days, larger and larger flocks of cowbirds have been building up in our backyard. Today, 200 came in to feed on the scattered grain. Yet this was the only record for this bird in the area. Not one of the 200 was seen by any of the twenty-two trained observers on the alert for any bird that flew. Apparently, all the cowbirds in the region—concentrated into this single flock—came directly to our yard and then left the area unseen.

DECEMBER · 30

THE DECLINE OF THE YEAR On this December afternoon, of all the twelve months only one day remains. I have seen the year shrink from seasons to months to weeks to days and now to hours. Its decline is nearly complete. Throughout this windy afternoon, I have been following old paths and walking among scenes that, during this slow contraction of time, I have known so well.

Out on the open sea meadows, there was no sound of clapper rail or sharp-tailed sparrow. Gusts, cold from the bay, swept over the flat land and gulls rode the long combers of the wind with wings swept back. The snow was largely gone. Only fragments remained like foam in troughs of the brown waves of the cordgrass. Overhead, hanging in the wind, on the alert for the stirring of mice, was the smallest of the falcons, a hardy sparrowhawk.

By Milburn Pond, snow still whitened the houses of the muskrats. The massed ferns that I first knew in early April as gray-green and furry fiddle-heads were now dry and cinnamon-brown. Along the shore, an edging of ice curtained the bottom where the sunfish scraped clean patches of pebbles for their spawning in the spring. The call of a downy woodpecker and the nasal repetition of a white-breasted nut-hatch came from the bare trees where warblers hunted and sang among the unfolding leaves of May and where warblers will sing again when the leaf buds on all the twigs have burst open in another spring.

Beside the swamp, the shaded path was a trail of snow. A mound of white formed a bower at the grass clump where the cottontail slumbered in its form that summer day. All up the slope of the weed lot, the goldenrod was gray and the grass was gray and the sky beyond was gray. Gray, too, was the swamp. I remembered it green from end to end. I remembered it gold and russet in the autumn. Now ashy-gray was the color of the fallen cattail leaves; ashy-gray was the pre-dominant hue of the lowland. The swamp stream was skimmed with ice in the middle and white with heavier, snow-covered ice at the sides. It looked like watered milk with cream at the edges. Riding a great gust, two-score red-wings made an immense wind-blown aerial leap over the swamp to the far phragmites. I heard no exultant "Oka-leee!" That was still far away. I heard only the sharp "Check!" of the months of cold.

I turned and walked up the snow-covered slope of the garden hillside. I leaned against the old Lincoln Tree, now holding within its galleries carpenter ants lost to the world in their winter slumber. I wandered among the dry stalks of the mallows and the sunflowers and the hollyhocks. I stood by the open space where, in summer days, ant

lions dug their pits and digger wasps made their tunnels. I
the tree trunk to which, like a tangle of gray-brown stri
vines of a morning glory clung. Its seeds had fallen to the g
were buried by the snow. I saw a Promethia cocoon riding a twig tip in the gusts and cattail fluff blown on the wind and scattering its tiny seeds down the length of the lowland. I turned toward home, into the cold wind, warm within. For everywhere around me in this snow-shrouded winter world there was evidence and promise of life and beauty to come.

DECEMBER · 31

A PLACE TO STAND Last sunset, last twilight, last stars of December. And so this year comes to an end, a year rich in the small, everyday events of the earth, as all years are for those who find delight in simple things. There is, in nature, a timelessness, a sturdy, undeviating endurance, that induces the conviction that here we have a place to stand. All around us are the inconstant and the uncertain. The institutions of men alter and disintegrate. Conditions of life change overnight. The thing we see today, tomorrow is not. But in the endless repetitions of nature—in the recurrence of spring, in the lush new growths that replace the old, in the coming of new birds to sing the ancient songs, in the continuity of life and the web of the living—here we find the solid foundation that, on this earth, underlies at once the past, the present and the future

INDEX